Recruiting Black Biology Majors into STEM Education Careers

This book addresses issues related to the recruitment, preparation, and retention of STEM teachers. Focusing on recruitment specifically, it explores the strategies used to introduce biology majors to the teaching profession, increase their interest in teaching, and support their transition into teaching. Taking the Transformative and Innovative Practices in STEM Education (TIPS) program as a case study, it draws upon a wide range of data sources to contextualize the experiences of program participants, including reflections from participants and program staff, pre- and post- surveys, focus groups, and annual interviews. The authors present insights about their decision-making and use the data to help create illustrative examples of the STEM majors of color who choose to pursue teaching and to explore why others decide not to pursue teaching. It foregrounds the importance of recruiting STEM teachers of color for urban districts, the role of culture and identity in the decision-making process, and the role played by professional development and mentoring. With emphasis on recruiting STEM majors at a Predominantly Black Institution (PBI), the book ultimately provides strategies for increasing collaboration across departments, supporting and mentoring students, and addressing cultural and institutional barriers that STEM majors face when transitioning into teacher education. As such, it will appeal to STEM education and teacher education scholars, as well as program directors, deans of Schools of Education, and deans of Schools of Science.

Salika A. Lawrence is Professor and Campbell Endowed Chair of Urban Education in the Department of Educational Leadership and Secondary Education at The College of New Jersey, USA.

Tabora A. Johnson is a Professor at Medgar Evers College, also Chair of the Multicultural Early Childhood and Elementary Education Department, CUNY, USA.

Chiyedza Small is Professor and Chairperson of the Biology Department at Medgar Evers College, CUNY, USA.

Routledge Research in STEM Education

The *Routledge Research in STEM Education* series is home to cutting-edge, upper-level scholarly studies and edited collections covering STEM Education.

Considering science, technology, engineering, and mathematics, texts address a broad range of topics including pedagogy, curriculum, policy, teacher education, and the promotion of diversity within STEM programmes.

Titles offer dynamic interventions into established subjects and innovative studies on emerging topics.

Queering STEM Culture in US Higher Education
Navigating Experiences of Exclusion in the Academy
Edited by Kelly J. Cross, Stephanie Farrell, and Bryce Hughes

Science and Technology Teacher Education in the Anthropocene
Addressing Challenges in the South and North
Edited by Miranda Rocksén, Elaosi Vhurumuku, Maria Svensson, Emmanuel Mushayikwa and Audrey Msimanga

Perspectives in Contemporary STEM Education Research
Research Methodology and Design
Edited by Thomas Delahunty and Máire Ní Ríordáin

Teaching Assistants, Inclusion and Special Educational Needs
International Perspectives on the Role of Paraprofessionals in Schools
Edited by Rob Webster & Anke A. de Boer

Invention Pedagogy—The Finnish Approach to Maker Education
Edited by Tiina Korhonen, Kaiju Kangas and Laura Salo

Visualisation and Epistemological Access to Mathematics Education in Southern Africa
Edited by Marc Schäfer

For more information about this series, please visit: www.routledge.com/Routledge-Research-in-STEM-Education/book-series/RRSTEM

Recruiting Black Biology Majors into STEM Education Careers

Journeys to Success

Salika A. Lawrence,
Tabora A. Johnson and
Chiyedza Small

NEW YORK AND LONDON

First published 2024
by Routledge
605 Third Avenue, New York, NY 10158

and by Routledge
4 Park Square, Milton Park, Abingdon, Oxon, OX14 4RN

Routledge is an imprint of the Taylor & Francis Group, an informa business

ISBN: 978-1-032-53949-2 (hbk)
ISBN: 978-1-032-53951-5 (pbk)
ISBN: 978-1-003-41445-2 (ebk)

DOI: 10.4324/9781003414452

Typeset in Times New Roman
by Apex CoVantage, LLC

Contents

Acknowledgments

We want to thank our families for their ongoing support and inspiration. We are also grateful to the students who have participated in the TIPS program over the past five years.

This project was enhanced exponentially by the special contributions of our STEM partners who designed and facilitated a range of professional development experiences for TIPS participants as well as the classroom teachers who provided invaluable mentoring.

The TIPS program would not have been possible without the generous financial support and ongoing partnership with our friends at the Voya Foundation.

Introduction

As former K-12 teachers who are currently higher education faculty in teacher education and science, we are concerned about the current challenges facing the teacher pipeline, specifically the lack of teacher diversity in low-income urban classrooms. In ***Recruiting Black Biology Majors into STEM Education Careers: Journeys to Success***, we discuss recruitment strategies for diversifying the teaching profession. There is a need for more Black and Latinx science, technology, engineering, and mathematics (STEM) teachers. There is also an urgent need for equity-minded individuals who can support the academic and social-emotional development of Black and Latinx students in low-income communities.

Several years ago, we launched a program to address these issues by leveraging resources across our departments at a Predominantly Black Institution (PBI). We believe that recruiting more Black and Latinx STEM teachers to work in high-need, under-resourced schools, especially urban schools, is an equity issue because many low-income schools are typically segregated with a high proportion of Black/African American and Latinx special education students as well as students learning English as a new language. Through our collaborative partnership, we designed and implemented a teacher recruitment program to increase the number of STEM teachers of color prepared to work in local urban schools. We've sought to increase our understanding of the recruitment strategies that appear to be most successful while examining the factors that influence Black STEM majors as they consider whether to become teachers.

Although our initial discussions as a team about the need for recruiting STEM teachers were informal, we noticed consistency with the high number of biology majors applying for a limited number of slots in the nursing program at the college, while simultaneously lamenting about the low number of students who choose STEM in the teacher preparation program in the School of Education. To further understand these patterns, we surveyed STEM majors from the School of Science, Health, and Technology (SSHT) to learn about their perspective on completing a teacher preparation program. The results of the survey showed that 47% of students considered becoming science teachers

DOI: 10.4324/9781003414452-1

and 60% would participate in a specialized program that will prepare them to become certified teachers. Most survey respondents cited "teacher salaries and working conditions" as obstacles that would prevent them from pursuing the teaching profession. The survey results were consistent with Han et al.'s (2018), who studied 60 different education systems across the globe and found that working conditions and teacher salary influenced students' desires to enter the teaching profession.

In 2018, we leveraged our access to STEM majors in the biology program and launched Transformative and Innovative Practices in STEM Education (TIPS) as a space to talk to STEM majors of color about teaching. Over time, the program became a space for faculty of color to mentor students, for peers to mentor each other, and for faculty to better understand the threats to STEM majors of color who seek to become teachers. In the past five years, we have tested various methods for introducing STEM majors of color to teaching as a viable career option and helped some of them become teachers.

In this book, we focus on the recruitment strategies used to recruit STEM majors into the TIPS program. We share lessons learned that we believe can help other institutions harness internal resources and capitalize on opportunities to recruit STEM majors of color into teaching, with a specific focus on Black STEM teachers. Throughout the book, we draw upon research supporting the need for more STEM teachers of color as well as strategies that have already been used to recruit them into teaching. We describe components of the TIPS program along with participants' stories, reflections and perspectives on program activities, and the impact of the program on the participants through their own words. The book also provides in-depth discussion of our methods for monitoring and improving the program. We share several resources and recommendations that can be used by institutions seeking to diversify their teaching pool by recruiting Black STEM majors.

Reference

Han, S. W., Borgonovi, F., & Guerriero, S. (2018). What motivates high school students to want to be teachers? The role of salary, working conditions, and societal evaluations about occupations in a comprehensive perspective. *American Educational Research Journal*, 55(1), 3–39.

1 The case for recruiting STEM teachers of color

Diversifying STEM education

It has been widely documented that science, technology, engineering, and mathematics (STEM) teachers remain a shortage area, particularly the low number of males and teachers of color in lower socioeconomic urban communities relative to student demographics (Beaudoin et al., 2013; Guha et al., 2017; Jones et al., 2016; Lewis & Toldson, 2013; Moritz & Weiss, 2018; National Comprehensive Center for Teacher Quality, n.d.). The low number of ethnically and racially diverse teachers compared to students in the classroom confirms the importance of recruiting more teachers who are representative of the students and communities they serve. The recruitment of Black teachers remains a dire global issue, especially in urban settings with proportionally more K-12 students from culturally, linguistically, and ethnically diverse backgrounds (Callender, 2020; Evans & Leonard, 2013), and, in order to address this issue, "teacher preparation programs are crucial to the recruitment and development of diverse teacher candidates" (Evans & Leonard, 2013, p. 2).

Although several recruitment initiatives have used targeted approaches and collaborative partnerships across institutions (Devari et al., 2017), recruitment and retention of STEM teachers remain a significant problem (Devari et al., 2017). Despite these initiatives, there remains an urgent call to bring more persons of color into the STEM fields, specifically in STEM education (Teo & Ke, 2014). Not addressing this issue can have a ripple effect throughout the K-20 STEM education pipeline, as K-12 students are disproportionately represented in STEM majors and few Black and Latinx teacher education candidates seek certification in STEM areas.

Some argue that recruiting STEM teachers is an issue of workforce development because K-12 students need deep engagement with STEM early (Milgrom-Elcott, 2016). If that's the case, supporting students in K-12 contexts should include resources and mentors to help foster students' interest in STEM. Elementary, middle, and high school students need teachers who can motivate, encourage, and inspire them because "the number of students that intend to join the STEM workforce decreases as they move from primary to secondary

DOI: 10.4324/9781003414452-2

grades and into postsecondary education systems" (Gandhi-Lee et al., 2017, p. 1). Initiatives seeking to recruit and prepare teachers of color aim to address inequities and systemic inequities in education, such as the disproportion of Black and Latinx students in science, technology, engineering, and math (Neally, 2022). These inequitable trends continue into college and education careers because students of color, particularly Black and Latinx STEM majors, do not want to pursue teaching careers because of the low pay, which places a pressure on them from family members to be financially secure compared to other better-paying careers (Chism, 2022; Lawrence et al., 2019; Neally, 2022). STEM majors of color who decide to become STEM teachers are influenced by their K-12 experiences in STEM and their perceptions of teaching as a highly valued career path (Lawrence et al., 2019; Neally, 2022). In fact, Lee et al. (2022) found that Black students in urban high schools reported that having a Black teacher significantly impacted the level of engagement and passion for STEM subjects. These results suggest that the limited pool of STEM teachers may be one factor contributing to the decrease in the number of students who choose to major in STEM subjects as they move up the grades.

Teacher educators need to think about innovative ways of recruiting prospective K-12 STEM teachers to fill current openings in teaching. Moreover, K-12 students need high-quality teachers who can help foster their content knowledge and support their interest in STEM subjects. Research shows that recruiting Black teachers to work with students along the K-12 pipeline can help strengthen STEM education and workforce initiatives seeking to diversify the pipeline (Chism, 2022). In our program we used targeted recruitment of high-performing STEM majors as a first step toward diversifying the field of teaching.

Representation and identification

Numerous strategies across the globe have sought to address the lack of representation and racialized and gendered inequities in the teaching profession through targeted recruitment efforts such as Black male teachers and Black STEM teachers (Callender, 2020; Chism, 2022; Green & Martin, 2018; The Journal of Blacks in Higher Education, 2019). This is important because students who are able to identify with their teacher's ethnic and gender background can be beneficial in academic areas where there is a lack of representation or negative stereotypes about some minority groups. UTeach, a teacher preparation program designed to recruit, prepare, and retain STEM teachers, will soon implement its research-based model in historically Black colleges and universities (HBCUs) to help diversify the teaching pipeline (The Journal of Blacks in Higher Education, 2020). Given the increase in the number of Black students attending PBIs and HBCUs (Green & Jor'dan, 2018), teacher recruitment initiatives that center these institutions are leading the field in the right direction by going to the source of a large pool of high-performing STEM majors because there is a shortage of qualified Black STEM teachers applying for teaching positions (D'Amico et al., 2017). Additionally, more structured institutionalized programs can help address the high attrition rates of people from historically marginalized groups in teacher preparation programs and in K-12 schools (Callender, 2020) by offering mentoring and supportive networks to address stereotype threats and challenge institutionalized racism and cultures that create barriers for people who may not have navigational capital and struggle to socialize into predominantly White spaces, where hiring practices and gatekeepers perpetuate the status quo (Callender, 2020; D'Amico et al., 2017; Yosso, 2005).

Education is highly racialized on many levels (D'Amico et al., 2017; Green & Jor'dan, 2018), including the teaching profession (D'Amico et al., 2017), with a disproportion of teachers to students of color in the classroom, the curriculum, and even the hiring practices, which can impact the percentage of Black teachers in today's classrooms. The short supply of Black teachers can be in part attributed to the fact that for decades because Black teachers only received jobs in large, low-income schools, or they were hired by Black principals compared to White principals, or in some districts "Blackness was negatively associated with receiving a job offer" (D'Amico et al., 2017, p. 27). The legacy of some biased hiring practices poses continued barriers for Black individuals seeking jobs in some school districts today.

Diversifying the teaching field is critical (Lewis & Toldson, 2013), especially in urban settings. In New York City, the location of the TIPS program, "more than 85% of the students are [students of color], [and] 60% of the teachers are white" (Rich, 2015, p. 6). Although the proportion of White and Latino/a teachers to students remained stable from 2002 to 2012, the proportion of African American teachers to students declined (Casey et al., 2015). According to a National Center for Education (NEA) report, Dilwort and Coleman (2014) noted, "a teaching force that represents the nation's racial, ethnic, and linguistic cultures and effectively incorporates this background and knowledge to enhance students' academic achievement is advantageous to the academic performance of students of all backgrounds, and for students of color specifically" (p. 1). Therefore, diversity in teaching is imperative because teachers of color are critical to school improvement efforts that aim to serve culturally and linguistically diverse students (Gist, 2017). Students of color benefit from having diverse teachers who share similar demographics (Albert Shanker Institute, 2015; Casey et al., 2015; Delpit, 2006; Delpit & Dowdy, 2008; Gay, 2018; Ladson-Billings, 2009; Vilson, 2016). This suggests that mentorship of students of color by teachers of color is important in STEM education and might make the difference as students progress in school. In fact, middle and high school students from all backgrounds prefer teachers of color to White teachers (Cherng & Halpin, 2016). For urban districts, which serve predominantly students of color, the issue is one of social justice and equity because under-resourced schools tend to have disparities in teacher quality (Cherng et al., 2022). Efforts to recruit and retain (Jones et al., 2016) persons of color in the STEM field could benefit from faculty mentorship, which has been shown to have positive outcomes. Griffin et al. (2010) interviewed faculty who mentored STEM students and found that mentorship is important for both undergraduate and graduate students. Their findings are consistent with other research (Patton, 2009; Patton & Harper, 2003) that highlights the significant impact of mentorship on STEM students' success.

To cultivate a skilled 21st century workforce that will meet science and engineering labor demands, institutions must place emphasis on improving STEM teaching so that K-12 students have high-quality teachers (Chism, 2022). There is a need for qualified teachers who are exceptionally committed to serving typically marginalized students in STEM because teachers play an essential role in ensuring students learn; they have great influence on the perceptions and experience of students in K–12 and beyond (Chism, 2022).

Severe consequences threaten the current educational system with the shortage of qualified STEM teachers. Experienced teachers are less prevalent in high-minority and urban schools (Cherng et al., 2022; Imazeki & Goe, 2009; Rahman et al., 2017; Rice, 2013). In addition, teacher certification and degrees in the subject taught varied significantly across student demographic groups. According to the 2015 NAEP survey (Rahman et al., 2017), 90% of eighth-grade mathematics teachers were certified by their states. Despite these numbers, schools with high percentages of minority students were less likely to be taught mathematics by certified mathematics teachers. The report showed that 75% of all eighth graders had a mathematics teacher with more than five years of teaching experience, but only 69% of those teachers were in schools with high-minority enrollment (Rahman et al., 2017; National Science Board, National Science Foundation, 2020).

The federal government projects that student enrollment will increase in public schools by 2024, with high-need school districts (i.e., schools with students at risk of educational failure or otherwise in need of special assistance and support) to increase most (Hussar & Bailey, 2016). Meanwhile, the number of teachers in public schools continues to fall as enrollment in teacher education programs show a steady decline, an issue that has been exacerbated by the global pandemic. The United States Bureau of Labor Statistics data have reported that there has been a net loss of 600,000 educators in public education since the onset of the COVID-19 pandemic in 2020 and that the ratio of hires to job openings in education has been reduced to 0.59 hires for each opening for the 2021–2022 school year, down from 1.54 in 2010 and 1.06 in 2016. This shortage of qualified teachers affects students' ability to learn as it creates an unstable teaching workforce that cannot be sustained due to high turnover (Ronfeldt et al., 2013; Jackson & Bruegmann, 2009; Kraft & Papay, 2014). How do we get more students to become STEM educators to supply a high-need demographic? To answer this question, we need to understand a few key factors.

First, the shortage of teachers exists due to multiple factors: (a) schools, especially high-poverty schools, are struggling to find qualified teachers because fewer students complete teacher preparation programs, high attrition, and high turnover; (b) low pay is making the profession unattractive; (c) challenging environments threaten teacher safety and create demoralizing attitudes; and (d) the almost complete lack of training, early career support, and professional development opportunities necessary to reinforce skills and ensure success. In high-poverty schools, this lack of training, support, and development is even more detrimental because teachers need to devote a great deal of time to instruction and have less time for professional development (García & Weiss, 2017).

Second, to solve this problem, schools and colleges must intervene early and pay attention when students express an interest in the teaching career, as well as address factors that stymie recruitment. Part of the answer must be to raise the number of students who enter the profession—particularly those interested in STEM teaching. Higher educational institutions and educators are critical to training and retaining students to meet these needs.

In a previous study, we examined the experiences of STEM majors to see how the TIPS program impacted STEM majors' decisions to become teachers (Lawrence et al., 2019). Since then, we've learned that similar results from Neally's work (2022) provide further evidence of the need for more recruitment-focused programs designed to identify and guide STEM majors into teaching. Neally (2022) conducted a study of STEM majors at ten universities to analyze the connection between their educational experiences and their decision to become STEM teachers. Research has confirmed that some of the experiences that positively influenced their decision to become teachers were personal connections with influential teachers who fostered a passion for STEM, a sense of belonging, and a support network in college, which included peer and faculty mentoring (Neally, 2022). Another consistent factor across studies is stereotype threat, specifically a lack of confidence in their knowledge and ability to teach (Lawrence et al., 2019; Neally, 2022). Content knowledge is a critical component of high-quality teaching (Evans & Leonard, 2013); however, even high-performing STEM majors experience stereotype threat in relation to their level of efficacy and self-perception (Lawrence et al., 2019; Neally, 2022).

Green and Martin (2018) compared the teacher preparation experiences of male pre-service teachers at a Predominantly Black Institution (PBI) and a Predominantly White Institution (PWI) and found that barriers to entry and matriculation for Black males include stereotypes about students of color, isolation, and limited culturally responsive practices. The findings from Green and Martin's (2018) research are consistent with other studies that purport the need for peer networks and faculty mentoring in postsecondary contexts. To understand how to recruit and support STEM majors transitioning to teaching, we need to understand their stories and perceptions of their experiences of postsecondary teacher preparation.

Strategies for recruiting STEM teachers of color

STEM teachers are considered a high-need area by the U.S. Department of Education (Bardelli & Ronfeldt, 2021), and recruiting diverse STEM teachers has become an urgent matter. Many districts and education programs are exploring innovative methods to meet the growing demand for STEM teachers. Across these programs, the common issue being addressed is the recruitment of teachers of color, particularly Black, Latina/o, and Indigenous persons, to better reflect the diversity in U.S. public schools (Madkins, 2011).

One of the strategies frequently used to address this issue includes funding from federal programs such as NOYCE (Morrell & Salomone, 2017). With financial support for scholarships and salary differentials from the NOYCE program, institutions of higher education can work with local partners to prepare high-performing STEM teachers from historically marginalized groups to serve low-income communities.

Our focus on teacher recruitment has helped to increase understanding of why some choose not to enter the profession and why some who have chosen it leave. A common factor reported by prospective teachers is the relatively low pay compared to other industries. In fact, teacher recruitment efforts, specifically those targeting Black and Latinx students, consistently face this barrier because

> the pay does not compete with other, more lucrative careers (especially when factoring in college debt), and teaching in these fields is not perceived as intellectually rigorous and worthy of high achievers. Even the U.S. Census does not include STEM teaching as a STEM career.
>
> (Milgrom-Elcott, 2016, p. 34)

For many Black and Latinx students who already feel the financial burden of high education costs, teaching is not perceived as a viable option for economic reprieve especially, if they have student loan debt. This trend suggests that teaching, specifically STEM teaching, is not prioritized and therefore can impact public perceptions of teaching and potential numbers in the pipeline and future pool of teacher candidates. Across the research, financial strain remains a common issue that can be addressed through scholarships that are embedded within the recruitment models (Jones & Cleaver, 2020). In Virginia, rural students enrolled in George Mason University's Rural and Diverse Student Scholars (RADSS) program receive a $5000 annual scholarship to diversify the teaching pipeline.

As the world transitions from the 21st to the 22nd century, recruiting and preparing STEM teachers is about workforce development and preparing teachers to support and foster students' interest and STEM literacy skills needed as the demands of STEM occupations are increasing (Milgrom-Elcott, 2016;

Whang-Sayson et al., 2017; Will, 2018). Most research on teacher recruitment preparation and retention has focused on the activities that lead to quantifiable outcomes and skills-based competencies, but more attention is needed to understand the experiences that can impact participants at different stages of the pipeline (Whang-Sayson et al., 2017).

One important variable is increasing awareness of teaching as a viable career option. Milgrom-Elcott (2016) noted that "in a world where tech superstars and venture capitalists make headlines moving millions of dollars, teaching STEM is not even on the radar as a career option for many talented young professionals" (p. 34). Despite changing attitude for those who decide not to enter teaching after completing a service-learning experience designed to introduce STEM majors to teaching, a survey of 173 respondents reported that more opportunities to engage in a variety of field-based experiences with students increased the likelihood of working in schools in the future or volunteering to teach (Whang-Sayson et al., 2017). Similarly, NOYCE Scholars at the University of Portland are "encouraged to take a service learning class in science or engineering, which introduces them to the education profession and provides them with field experience in a high-need classroom" (Morrell & Salomone, 2017, p. 17).

Recruitment programs have also provided invaluable opportunities for STEM majors to hone their skills and gain experience working with diverse learners. STEM teacher recruits in the Teacher TRAC program gain practical experience during early field internships, where they tutor students with special learning needs, such as the developmentally disabled, English language learners, and children with autism (Parsons, 2013). Not only does this program allow prospective STEM teachers to hone their skills teaching in critical areas such as science and math, but they also learn how to work with a range of diverse learners. STEM majors also get to increase their teaching skills by participating as teacher assistants in a summer middle school academy where they work alongside a middle school teacher and a professor to teach inquiry-based activities for middle school students (Parsons, 2013).

A review of 30 years of research shows successful recruitment efforts have included alternate routes and apprenticeship models such as residency programs (Madkins, 2011). Although all alternate route programs rely heavily on partnerships, the models vary (Douglas & Khandaker, 2015; Grossman & Loeb, 2010). Alternate route programs have grown exponentially since the 1980s (Madkins, 2011). Many of these programs have been "effective in increasing the number of Blacks in the teaching force" (Madkins, 2011, p. 422), many of whom have been older, nontraditional students. Regardless of the type of program, this research suggests that recruitment should focus on high-quality candidates, on-the-job coaching and feedback for novices, and restructuring coursework around best practices in the field, which are promising practices (Grossman & Loeb, 2010).

Several historical factors precipitated the current loss of Black teachers in urban schools, such as the lack of teaching opportunity in desegregated schools as Black students were bussed to white schools, more employment opportunities in other industries, particularly STEM fields, such as medicine, which are more financially advantageous, and Black students in the pipeline who are academically underprepared for postsecondary education and career gatekeeping metrics such as SAT and teacher licensure exams (Madkins, 2011). Numerous federal and local initiatives have sought to increase recruitment and retention of Black teachers so the staff is more representative of students, particularly in urban districts, but the disparities between teacher and student demographics remain (Albert Shanker Institute, 2015; Madkins, 2011).

Some districts have sought to recruit and prepare their own STEM teachers to address the shortages (Will, 2018). Many teacher education programs have partnered to offer alternative pathways into teaching. Collaborative partnerships are a core aspect of STEM recruitment and preparation. Milgrom-Elcott (2016) reported that "partnership between the Colorado School of Mines, the University of Northern Colorado and the Denver Teacher Residency provides a way for engineering graduates to be trained and certified and placed in STEM classrooms" (p. 35). Although enrollment has declined in teacher education programs, endorsement certification has helped to increase the number of teachers certified to teach in high-need areas such as special education, STEM, and bilingual/ESL (Bardelli & Ronfeldt, 2021).

For over two decades, teacher education programs have partnered with STEM professionals to recruit and prepare STEM teachers through the federally funded Robert Noyce Teacher Scholarship Program (NOYCE) (Morrell & Salomone, 2017). These programs have increased the pool of STEM teachers in urban schools as program participants commit to teaching in high-need schools. While in the program NOYCE, scholars participate in a Saturday Academy internship working with students in grades 2–12. The classes in the Academy are taught by community professionals, and the NOYCE scholars have the opportunity to design and teach hands-on STEM lessons (Morrell & Salomone, 2017).

There is an ongoing need to diversify the current teaching profession and pipeline, particularly in STEM areas. Despite its controversies (Curran, 2017), Teach for America (TFA) has been recognized for its ability to recruit "academically strong and motivated young people who would otherwise not consider teaching, especially in high-poverty schools" (Donaldson & Johnson, 2011, p. 48). TFA has been viewed as the solution for addressing staffing in shortage areas (Curran, 2017). The TFA program has expanded internationally, like in India, for example. Teachers in the Teach for India program face challenges linked to diversity because the teachers recruited to teach in high-needs, impoverished and overcrowded schools do not share language or other cultural backgrounds as the students they are teaching (Joseph, 2016). Diversity and cultural responsiveness have been an issue (Perry, 2015), particularly because of the disconnect between the program and the communities they serve. TFA has attempted to increase recruitment of teachers who identify as Black, Indigenous, and Latinx (Perry, 2015). It has attempted to partner with Black and Latinx sororities and fraternities to increase recruitment of teachers of color (Munoz et al., 2019), but counter-narrative analysis of TFA's practices shows that

> TFA does not align with the missions of Black and Latinx F&S [fraternities and sororities]. By engaging specific property functions of whiteness, we have demonstrated that Black and Latinx F&S' partnerships with TFA are inconsistent with their empowerment mission that focuses on lifting Communities of Color via philanthropy and service. Instead, TFA maintains the whiteness as property status quo and thus perpetuates interest convergence within Black and Latinx F&S that were originally formed to survive and resist domination.
>
> (p. 68)

Utilizing networks of Black and Latinx sororities and fraternities is a viable approach because it leverages community networks as resources that can serve schools in communities of color (Munoz et al., 2019; Yosso, 2005). This asset-based approach draws upon critical race theory to transform oppressive and marginalized spaces in urban communities by empowering the members of those communities as insiders (Munoz et al., 2019; Yosso, 2005). To do so, counter-narratives (Munoz et al., 2019) help to prioritize performing teachers of color in communities of color to dismantle systems and change biases and perceptions of students of color. Munoz et al. (2019) urge that programs use a critical race theory and cultural wealth model when establishing partnerships for teacher recruitment and preparation.

Previous research shows that advertising and disseminating information about the program should include methods such as "Facebook posts, website listings, flyers, and word of mouth" to help people consider teaching as a career (Morrell & Salomone, 2017). After interviews with STEM faculty across disciplines, Gandhi-Lee et al. (2017) report that STEM recruitment and retention are influenced by

- an early interest in STEM-related careers based on interest in specific subject areas and academic preparation to build confidence and efficacy in STEM subjects; and
- student-to-student and faculty–student interactions inside and outside of the classroom through mentorships, collaborative research, and networking opportunities.

While faculty-led experiences create learning opportunities and environments that foster interest in STEM fields, similar introductions to and preparation for STEM teaching are limited. STEM faculty can be instrumental gatekeepers for fostering the STEM education pipeline, and therefore it is important to increase STEM faculty awareness of and stake in STEM education by supporting recruitment through outreach (Gandhi-Lee et al., 2017) and increasing retention by serving as role models (Hutton, 2019).

Kunz et al. (2020) report that some other factors that influence STEM teacher recruitment include their motivation to teach: intrinsic factors such as teaching content they enjoy or altruism by serving and contributing to the broader society; and external factors, namely job security and salary. Altruism, particularly among millennials, is one of the most common factors motivating STEM majors to enter teaching (Kunz et al., 2020). Although prior to a five-day job shadowing experience, 40 STEM majors reported their motivation to complete the job shadowing was intrinsic—specifically to learn skills as a prospective teacher in the content and to better understand the profession—none of them indicated an altruistic motive to participate in the program until after the job shadowing because they were able to work and bond with high school students (Kunz et al., 2020).

Hubbard et al. (2015) explained that the Talented Teachers in Training for Texas (T4) program was designed for STEM majors with an interest in teaching but who were not pursuing teacher certification at the time they entered the program. Over time, those who commit to teaching in a high-need district can receive mentoring, a scholarship, and an in-service induction program when they start teaching. Recruitment efforts for T4 included university faculty and other students visiting STEM classrooms to speak about teaching as a career (Hubbard et al., 2015). The program sought to increase awareness of STEM teaching by using outreach to provide STEM majors with exposure to this career choice through a range of experiences, including job shadowing (Hubbard et al., 2015). Job shadowing provided the "T4 recruits, [with] a realistic window into the STEM teaching profession and, at the same time, develop an authentic working relationship with an accomplished science or mathematics teacher" (Hubbard et al., 2015, p. 71). STEM majors can also increase their content knowledge and communication skills by working with middle and high schools on projects (Ferrara et al., 2018).

The Teach Tomorrow in Oakland (TTO) pipeline program provided a pipeline for Latinx, African American, and Asian undergraduates, recent graduates, and career changers interested in teaching (Rogers-Ard et al., 2012). Additionally, role models can help support a sense of belonging and the social and emotional skills needed to persist through STEM courses while also helping to increase academic and content knowledge (Hutton, 2019). K-12 teachers and higher education faculty play a critical role in fostering students' interest in STEM education. According to Hutton (2019), "even if students do not intend to pursue a STEM career, they still need to be STEM literate to participate in the 21st century economy and workforce" (p. 16).

Factors impacting recruitment of STEM majors of color

As noted earlier, several factors influence students' educational and professional interests in STEM at the secondary level, including limited exposure to high-quality STEM education in K-12 contexts as they progress in school. It is also important to recognize the contribution of cultural background, socioeconomic status, gender, ethnicity, parental education, students' achievement in elementary and secondary STEM courses, as well as parental expectations (Svoboda et al., 2016; Gemici et al., 2014; Galaviz, 2020). Socio-cultural factors may be just as important as access to the STEM education curriculum. There is a need for more research to look more closely at high-performing students of color who have access to high-quality K-12 STEM education but ultimately do not enter STEM fields. The few existing studies highlight the impact of family culture and thus reinforce the need to focus our study on African American and Afro-Caribbean undergraduates.

We saw a repeated theme in surveys, feedback, and discussions with our participants in the TIPS program: significant role of family dynamics and influence. This led us to the following questions: What makes families so important to students who are making their own choices? What are some of the tensions participants are facing, particularly the tensions between what I see of myself and what I can do, but the fear? Are there specific cultural/historical perspectives and beliefs? Below, we capture a few of their stories to show the individuals who represent some of the trends we noticed in our participants.

African epistemology centers on the family as its core; thus, it is no surprise that for our participants, families play a significant role in their career choices. During the end-of-year interviews with Cohort 1, we learned that for our participants, who were all Black females, their family opinions about their career choices often superseded their own desires. Our participants were more often than not directed to apply to the medical field because of the prestige that it offers along with the salary that families believe that doctors make. We debunked the myth that all doctors are high-income earners immediately by comparing a first-year teacher salary in New York City with that of a nurse and a first-year medical doctor. Participants were surprised when they saw that teachers with more credits and higher degrees made more than nurses and even some doctors.

Jazmin, from our first cohort of students, had a passion for math and wanted to teach the subject. Whenever she spoke of the subject, her excitement was visible. She explained during our conversations and end-of-year interview that while we wanted to teach, her family dissuaded her from becoming a teacher and pushed her to enter the medical field. Their influence seems to override her own desires and excitement to teach math. While family members may be well intended, we found the desire to push students to become medical doctors was powerful and at times overshadowed the student's personal career decisions, derailing them from choosing or viewing teaching as a viable option. Familial expectations and cultural Influences in Pursuit of STEM for Black students of various ethnicities, such as African American, Caribbean-American, college students from across the African continent, seem to be a barrier to Black STEM majors becoming teachers. There is a tension between what students see for themselves and can do, but fear exists when families remove personal career choices as an option.

Family habitus in K-12 STEM education

In many studies, researchers have identified the role families play in developing student interest and achievement in science (Archer et al., 2012, 2015; Claussen & Osborne, 2012; Dabney et al., 2013; DeWitt et al., 2013; Adamuti-Trache & Andres, 2008; Aschbacher et al., 2010, 2014; Gilmartin et al., 2006). Some researchers have highlighted how family involvement can influence the early interest students show in STEM and how family capital can impact the levels at which a student participates in STEM activities (Dabney et al., 2013; Archer et al., 2014; Rodrigues et al., 2011; Mau, 2003; Atherton et al., 2009). There is little research specifically on the impact of family habitus (i.e. the way an individual of a particular background perceives and reacts to the social world) as it pertains to underrepresented STEM college students choosing majors and careers.

Parents play a critical role as an early influence who can best motivate adolescents to take high school courses that will better prepare them for STEM careers. To make lasting changes that may affect the outcomes for ethnic minority students, it is worth examining how that influence affects education trajectory and career decision-making (Hill & Tyson, 2009; Hyde et al., 2006; Galaviz, 2020). In a study of Australian high school students examining what drives the educational and occupational aspirations of students, Gemici et al. (2014) found that parental influence and academics were two of the strongest predictors of academic performance. This study also found "the most influential factors for students' aspirations for completing year 12 include their academic performance, immigration background, and whether their parents expected them to go to university." They further wrote that "students whose parents want them to attend university are four times more likely to complete Year 12 and 11 times more likely to plan to attend university compared with those whose parents expect them to choose a non-university pathway." Overall, our investigation illustrates a significant, though less-emphasized, consideration in higher education in the United States—that is, how family habitus can determine academic outcomes. The question is, "How can we successfully leverage this knowledge to influence the outcomes of African American and minority students in pursuit of STEM disciplines?"

In *The Colour of Class*, a study conducted in England, Rollock et al. (2015) focused on intersections between race, class, and parental involvement in their children's education. Specifically, they examined Black, middle-class Caribbean families who are successful "achievers and professionals" and their societal positioning and experiences with multiple identities as Black, British, and middle-class professionals in majority "white spaces" at work and school. Parents used deliberate efforts and strategies to ensure their children achieved successful educational outcomes while navigating racism, class, and ethnic background and maintaining Black cultural identities. This research further underscores the complexities around educating children and the role parents play; it offers lessons for college students whose cultures and ethnic backgrounds undergird their academic and career choices. Can "parental capital" improve outcomes for college students in STEM?

The work of Harackiewicz et al. (2012) explored Eccles's 2009 expectancy-value theory, establishing how "parental influence" can be a resource in directing how adolescents select courses and persist later toward STEM careers. The expectancy-value theory (Eccles, 2009) postulates that people are more likely to choose a challenging task if they value the task as useful (utility value), relevant, and expect that they can succeed at the task. Parental beliefs and behaviors can shape their children's choices; for example, if parents place value in their children taking advanced STEM courses in high school, children may be motivated to place similar value in those courses and in STEM career choices (Eccles, 2009).

Familial expectations and cultural influences in pursuit of STEM for African American and Caribbean-American college students

The educational experiences of African American and Afro-Caribbean American college students in America traverse racial, socioeconomic, cultural, and at the very least academic factors that impact career choice, success, and ultimately upward mobility. STEM research on African American students will often "merge native-born and foreign-born blacks as a single demographic . . . without examining the cultural differences among the two groups." However similar ethnically and racially, these two groups "differ in their perception of post-secondary education and their response to adversity (academic attainment) in STEM" (Leggett-Robinson, 2017; Griffin et al., 2012). Foreign-born students account for over 27% of the Black student population at selective colleges across the United States (Massey et al., 2007; Leggett-Robinson et al., 2017). Ogbu and Simons (1998) argue that the cultural differences that impact the performance gaps between the two groups can be explained by cultural distinction theory, which suggests

> that foreign-born blacks possess cultural values and experiences that are distinctive compared to those of native-born blacks, who have an ancestral history of slavery and post-slavery experiences. The differences may enhance socioeconomic and educational attainment of foreign-born blacks,

> who were socialized in a society where they are the racial majority, thus potentially creating an achievement gap between foreign-born and native-born blacks.
>
> (Ogbu & Simons, 1998; Leggett-Robinson, 2017, p. 5)

Though we do not delve deeper into the differences in the cultural backgrounds of our participants, we acknowledge that its importance warrants further research.

The pool of high-performing STEM majors

Although the initial focus of the TIPS program has been on biology, there is a potential pool across subject areas (Table 1.1). However, one potential threat hindering STEM majors of color is the stereotype of what characterizes a high-performing STEM major.

Harackiewicz et al. (2012) tested whether they could "influence adolescents' motivation to take science and math courses by providing information about utility value to parents." In a study of tenth and eleventh graders and their parents, the researchers provided a STEM intervention "intended to influence parents' values and interactions with their children and ultimately influenced their children's course choices" (Harackiewicz et al., 2012, p. 1). This research showed that if parents were persuaded of the utility value of math and science and could communicate that value to their children, it led to children taking more math and science courses in high school. In particular, the mothers' perception of the value of STEM courses "promoted parent-child conversations about the value of STEM courses, and increased the number of STEM courses adolescents took during the last two years of high school" (Harackiewicz

Table 1.1 Total student enrollment, School of Science Health and Technology

Name of STEM departments	*BIO (BS)*	*CHEM AND ENVS (BS)*	*MTH (BS)*
Total number of students in each STEM department	1,171	38	41
Current number of UR students in each STEM department	1,102	34	39
Total number of students in each STEM department with GPA >3.0	474	17	21
Total number of UR students in each STEM department with GPA >3.0	432	15	19
Total number of students in each STEM department with GPA >3.5	315	13	14
Total number of UR students in each STEM department with GPA >3.5	280	11	13

et al., 2012, p. 6). Harackiewicz et al. (2012) identify parents as "an untapped resource" for STEM intervention programs, but can that influence have limiting effects and create barriers for students' choices in college as it pertains to STEM teaching? We explore how family expectations serve as a source of strength for underrepresented STEM students—by validating STEM capability and self-efficacy—even as these expectations can reinforce negative perceptions of STEM teaching.

Abe and Chikoko (2020) studied factors that influence the career decisions of STEM students at a university in South Africa. They found that "the role played by family in the career decisions of students was more significant than monetary influences." Forty-five percent of students said that their families were "very influential" in their decision to pursue a career in STEM, while 20% of participants said teachers played a "significant role" in their career decision-making. This study also revealed that students cited "the need to support family" as another factor that presented an "unexpected sub-theme that emerged from family influence on career decision-making in this study."

Culture and stereotype threats

Beyond bias and perceptions around academic performance, threats also include ingrained cultural ideas that are transmitted across generations in subtle and over ways. Most of our participants are nontraditional students, immigrants, and low-income students from low-income communities who overcame academic challenges to be high-performing STEM majors at college. Therefore, we sought to cultivate a community for STEM majors where they could get support and access to resources through the TIPS network. From a student's perspective navigating institutions of higher education can be challenging. Even more perplexing are the demarcations found across departments. With this in mind, we adopted a cultural capital framework for the program. We believe that culture goes beyond ethnicity and race to include "characteristics and forms of social histories and identities . . . [and] behaviors and values that are learned, shared, and exhibited by a group of people" (Yosso, 2005, p. 75). Furthermore, embedded in the TIPS program is an asset view of cultural wealth, specifically social capital—networks and resources. Therefore, "social capital implies that people well-equipped with social resources—in the sense of their social network and the resources of others they can call upon—better succeed in attaining their goals" (Lancee, 2012, p. 17).

Given the benefits of a community as a source of wealth (Yosso, 2005), mentoring offered through networks should be viewed as multilayered and essential because mentees benefit from peers, teachers, and experts. A social capital framework, which "takes the form of social ties, obligations, connections, and networks" (Samuelson & Litzler, 2016, p. 95), helps to centralize practices around the relationships and experiences of the participants to help affirm the individual, their values, and their cultures. For example, one type of community in academic contexts that African American and Latinx students found beneficial is the aspirational capital community, which provides a space for discussions about their future, aspirational goals, and future careers. This context helped students explore "the idea that their future academic attainment and occupation are not necessarily linked to their parents' current occupational status" (Samuelson & Litzler, 2016, p. 97).

There is a dominant narrative about Black/African American and Latinx students compared to their White peers, and this negative portrayal is damaging. This narrative includes the underrepresented students who are less motivated, interested, and capable of excelling academically compared to White students. Effective teacher recruitment and retention strategies to support students of color, particularly males, need to focus on helping students of color "overcome negative societal perceptions and racist stereotypes" (Wallace & Gagen, 2020, p. 420).

Ogbu (1992) compared ethnic minority groups and pointed out that it was cultural factors, not genetic differences, that influenced academic achievement. His idea suggests that academic performance is within student control, and he discounts the structural barriers that pose challenges. His research suggested that educational challenges for African American students were due to "community forces . . . [that] serve to differentiate minority groups facing similar barriers in society at large and in schools; and the options created by the community forces allow choices of action that resulted in individual differences in schooling outcomes" (p. 287). He also suggested that cultural models provide minority groups with access to understanding the world. When looking within minority groups, Ogbu (1992) suggested that immigrant or voluntary minorities who moved to the United States on a voluntary basis have greater economic outcomes and academic achievement in U.S. society than nonimmigrant or involuntary minority groups resulting from slavery, conquest, or colonization.

Social capital

Research examining the educational experiences and trajectory of Black students from elementary to post-secondary levels, particularly in STEM, describes their experiences through expectancy-value theory, social capital theory, and social cognitive career theory (Nugent et al., 2015). Leggett-Robinson (2017) revealed that foreign-born Black students had higher grade-point averages and more success in graduating than their native-born Black counterparts in the Science, Technology and Engineering Programs (STEP). Although African American groups in STEM have made significant progress, underrepresentation still exists, and it will continue to impact critical areas in education and the STEM workforce. Therefore, we must understand the barriers that may limit STEM students' opportunities and choices.

TIPS is an asset-based framework informed by social and cultural wealth theories (Yosso, 2005) because we used these ideas to design a supporting program to help participants as they transition into teaching. At the core of the program are the interrelationships formed to help center the experiences of the participants. Forming these kinds of social networks (Yosso, 2005) helps facilitate the exchange and sharing of knowledge, skills, values, and competencies, which can inform key identity shifts in participants who receive exposure that they might not have otherwise experienced. Participants are viewed as fully developed beings who strengthen the micro and macro community and can help to dismantle inequitable structures. The personal interactions between individuals in the group help to reinforce the community by leveraging social resources that come from participating in the group (Ali-Hassan, 2009; Durlauf & Fafchamps, 2005). We found this community-building to be a critical resource for STEM majors transitioning to teaching. Despite Gandhi-Lee et al.'s (2017) argument that some students may not have familial STEM role models, which can create a barrier to their pursuit of STEM, many of the TIPS participants reported familial pressure to be STEM majors, particularly to pursue a career in medicine. This same familial pressure, which we will discuss further in other sections of this book, was also a barrier for many who considered teaching.

The program is guided by the premise that access to resources can help STEM majors of color who are navigating higher education as first-generation students or nontraditional students and provide them with a broader perspective of opportunities available to them. For some individuals, increasing their knowledge of mainstream, dominant social systems can help liberate them from a perpetual cycle that often keeps marginalized groups in the lower class. Yosso (2005) reframes Coleman's (1988) social capital theory to focus on networks and ways to leverage local knowledge in communities. From this perspective, the emphasis is on the role individuals play in networks and groups and the ways in which humans interact within institutional contexts to create systems that benefit members of the group. Through these interactions, the group builds trust and finds ways to share, build upon, and advance the assets of its members. The personal interactions between individuals in the group help to reinforce the community by leveraging social resources that come from participating in the group (Ali-Hassan, 2009; Durlauf & Fafchamps, 2005).

A social capital framework can "take the form of social ties, obligations, connections, and networks" (Samuelson & Litzler, 2016, p. 95). It helps to centralize practices around the relationships and experiences of the participants to help affirm the individual, their values, and their cultures. For example, in an academic context, one type of community that African American and Latinx students found beneficial is the aspirational capital community, which provides a space for discussions about their future, aspirational goals, and careers. It helped students explore "the idea that their future academic attainment and occupation are not necessarily linked to their parents' current occupational status" (Samuelson & Litzler, 2016, p. 97).

Another community is navigational capital, which "denotes skills that help students maneuver through educational institutions with dominant cultural norms" (Samuelson & Litzler, 2016, p. 97). For many STEM majors of color, their academic success is part of their confidence. Samuelson and Litzler (2016) found that Latinx and African American engineering students believed strongly that their academic performance provided them with the confidence needed to persist and navigate higher education.

Structures that affect the shrinking pool of STEM teachers of color

The lack of diversity in the teaching profession is magnified when one compares the number of students of color to the number of teachers of color in American public schools. This disparity confirms the ongoing need for more teachers of color (Wallace & Gagen, 2020). Within the current discussion on diversifying the teaching profession—specifically recruiting, preparing, and supporting STEM teachers in urban, low-income schools (Albert Shanker Institute, 2015; Lewis & Toldson, 2013)—TIPS builds upon previous research to broaden the conversation, especially ways to recruit and support STEM majors of color.

To meet the demand for STEM teachers, institutions have included innovative methods such as targeted recruitment of students of color, alternate route programs for career changers, community-based models that prepare community residents to teach, and future teacher clubs at middle and high schools (Hill & Gillette, 2005; Rogers-Ard et al., 2012; Wallace & Gagen, 2020). Ongoing structural barriers continue to prevent some STEM majors of color from entering the teaching profession. For recruiting and placing African American male teachers, Rogers-Ard et al. (2012) identified several barriers, the largest of which was an 80% failure rate on the state certification exam. Out of a prospective pool of ten African American males accepted into their recruitment program, only one candidate was able to receive placement as a math teacher (Rogers-Ard et al., 2012).

Another challenge is the shrinking pool of potential candidates, which might be linked to general perceptions of the teaching profession. The U.S. Department of Education (2019) reported a 19% decline in degrees in education from 2006 to 2017, as fewer people are pursuing degrees in the field of education. The perception of teachers, myths about the teaching profession, and the portrayal of urban education are ongoing challenges. Perhaps because public funds are used to support schools, public critiques have led to negative narratives about teachers and public education that may have contributed to the current teacher shortage.

Trends in social issues and public crises tend to sway beliefs and public opinion, and the media plays a crucial role in public perception (Quinn, 2020). Teachers in the United States, and specifically, public education, have long been under the microscope (Quinn, 2020). According to Quinn (2020), publicly disseminated reports about student performance on high-stakes tests and teacher evaluation data have spurred discussions across sectors outside of education and debates about teacher quality, performance, and preparation.

Another factor is the cultural and institutional capital needed for students to navigate academic spaces to support their transition from STEM major to STEM educator. When developing the TIPS program, we used an asset-based view of cultural wealth; we emphasized the social capital required for a STEM major to identify supports for transitioning into teaching. From this perspective, "social capital implies that people well-equipped with social resources—in the sense of their social network and the resources of others they can call upon—better succeed in attaining their goals" (Lancee, 2012, p. 17). These kinds of social capital are often displayed by people of color in academic spaces.

Teachers of color tend to be cultural brokers for their students (Villegas & Irvine, 2010), and that role is no different for faculty of color in higher education spaces. Gist (2017) found that pre-service teachers of color benefited from having faculty of color and perceived that a professor who can "cultivate cultural and linguistic affirmation is vital" (p. 943). Black and Latinx teacher candidates expressed pride and affirmation from their experiences with faculty.

For many pre-service teachers of color, interactions with faculty of color help support them through personal and familial tensions about their career choice because the faculty represented the possibilities along with "professional success in prestigious professions" (Gist, 2017, p. 939). Pursuing teaching presents personal conflicts such as "contemplating clipping personal ties to family or resisting negative evaluations of cultural and linguistic affiliations . . . [as they] fought to stay on track in attaining their professional goals" (p. 940). As one participant in Gist's (2017) research noted, the mentorship from faculty of color helped to address "the internal struggle to racially/ethnically and culturally distance herself to attain better professional opportunities for her future" (p. 940).

Although previous frameworks attempt to critique social structures (see Bourdieu & Passeron, 1977; Ogbu, 1992; Ogbu & Simons, 1998), they perpetuate deficit views of minority groups by overlooking the cultural assets students of color bring with them to school. A shift in perspective must consider interpersonal networks because "social antagonisms and divisions existing in the wider society operate to problematize (if not undermine) minority children's access to opportunities and resources that are, by and large, taken-for-granted products of middle-class family, community, and school networks" (Stanton-Salazar, 1997, p. 3). These resources can include local, community-based resources often overlooked, namely libraries, places of worship, community and childcare centers, and other institutions within neighborhoods that may be underutilized assets that can help support educators (Green, 2017; Small, 2006).

Addressing the deficit thinking presented by previous perspectives means theorizing that communities of color bring assets—community and cultural capital wealth—with them into academic spaces that can positively impact the schooling of students of color (Green, 2017; Kinney, 2015; Yosso, 2005).

One asset Yosso (2005) identifies is social capital, which includes the networks and community resources, including peers and social contacts, utilized to support higher education, employment, and other pursuits. Social capital in the form of mentoring (Gist, 2017) helps to motivate students of color, particularly STEM majors, so they persist through academic spaces that pose structural barriers (Samuelson & Litzler, 2016). Social networks that include institutional agents can offer mentorship and access to resources and opportunities (Gist, 2017; Stanton-Salazar, 1997; Yosso, 2005).

In most post-secondary institutions, "traditionally disciplinary faculty in IHE [Institutions of Higher Education] STEM departments have contributed to K-12 education by teaching content-specific courses for pre-service and in-service mathematics and science teachers at the IHE, rather than participating directly with teachers in K-12 schools" (Moyer-Packenham et al., 2009, p. 1). Moving away from these silos, and toward a community-of-practice model (Sack et al., 2016), both STEM faculty and education faculty will provide mentoring and on-site coaching to program participants. Mentoring has a positive impact on academic performance, behavior, and social skills and helps students transition and be successful in college programs (DuBois et al., 2011). Additionally, students who received mentoring were more likely to persist in college, gained greater knowledge of campus resources, and earned higher grades than students in comparison groups (Bettinger & Baker, 2011). Preparation for STEM teachers tends to be discipline-specific because math and science teachers generally use different approaches to reasoning in each subject area (Wasserman & Rossi, 2015). Collaborative partnerships between higher education departments, K-12 local education agencies, and private partners have positive benefits for teacher preparation and professional development and frequently lead to improvements in student outcomes (Moyer-Packenham et al., 2009).

Who are the TIPS participants?

The majority of the TIPS participants are high-performing STEM majors. Many of our participants have also overcome academic barriers and preparation due to lower test scores and reading levels. Through the TIPS program, we seek to bring their voices to the center of discussion, so they are no longer overlooked as potential candidates for teaching positions.

A 2022 biology department survey reported that 52.3% of students aspired to pursue careers in healthcare and medical fields, 22% in nursing, 11.7% as a medical doctor, and 1.3 % as a science teacher, an 18% decline from before the COVID-19 pandemic. However, these numbers do not reflect the limited capacity of most nursing and medical programs in the United States, particularly at our college, to accommodate large numbers of students. This leads to high competition for those accepted, leaving many students to figure out alternatives if they are not accepted into the programs. More often than not, biology students will keep the job they did before obtaining their degrees or resort to retail jobs that do not require the use and skills of their degrees.

When planning, we realized that a recruitment plan for TIPS had to first acknowledge the factors that prevented students from considering a teaching career, which primarily include a need for more information about the profession and negative perceptions around salary, work responsibilities, and managing classrooms. Recruitment efforts for the TIPS program focused on recognizing an available pool of high-achieving STEM students who could benefit from a comprehensive program designed to introduce them to a teaching career. Seventy percent of participants strongly agreed or agreed that the program increased their interest in teaching.

TIPS is housed within a PBI located in a large urban community; thus, our participants over the course of five years have primarily been Black with multiple ethnicities, highlighting the African Diaspora (see Table 1.2). We use the term Black for consistency to reference persons of African descent throughout this research; there are times when we may use African descent interchangeably. Participants in our study primarily identify as Black, with the largest ethnic group being African Caribbean, a designation used for students who were born in the Caribbean or who are first-generation immigrants, with one or more parents from the Caribbean. All program participants have been biology majors. Some students also had minors in mathematics, computer science, environmental science, criminal justice, and psychology.

The project was started with the intention to "grow our own garden." Our diverse pool of Black students allowed us to center the voice of Black STEM majors, who bring rich, valuable experiences, thoughts, and innovations to the field. We posit they've always been here, but programs have overlooked them in the past. We chose to focus on Black STEM majors because very few research studies focus exclusively on Black students. While there are other persons of color in our program, the majority are Black and have given greater perspective to some of the barriers Black students face in becoming STEM educators, including, but not limited to, the lack of STEM preparation, literacy readiness and reading levels, and exposure to STEM and AP courses. Many urban school districts have critical teacher shortages, which can have implications for effective teaching and student success. Through our work, we've found that Black STEM majors, from lower socioeconomic backgrounds, may enter college with less exposure to STEM subjects and experiences than their peers. If we want to diversify the STEM field and increase the number of Black STEM teachers, we must identify the core issues that students face in completing STEM degrees at the college level and then address and remove obstacles. The purpose of our signaling Black students highlights the need to focus on them if we want to diversify teaching.

Table 1.2 Demographics of TIPS participants

Year	*# (start)*	*# (end)*	*Demographics*	*Age range*	*Ethnicity*	*STEM field*	*Classification*
1	12	4	90% female, 10% male, 100% African American/Black, 80% foreign-born: Jamaica, Trinidad and Tobago, Haiti	24–29	4 African-Caribbean	Biology	2 juniors 2 seniors 2 nontraditional students 4 working part-time 1 parent 1 after-school tutor in a public school
2	10	5	90% female, 10% male, 100% African American/Black, 80% foreign-born: Jamaica, Trinidad and Tobago, Haiti and a continental African student born in Morocco	22–37	1 Moroccan 2 African- Caribbean 1 Black	Biology Math	5 seniors, all nontraditional students 1 parent
3	10	5	90% female, 10% male, 100% African American/Black 90% foreign-born: Nigeria, Canada, Trinidad, and Tobago, Jamaica, Morocco	21–33	2 Nigerian 1 Moroccan 1 African-Caribbean 1 Black	Biology Environmental science	4 seniors 1 junior 1 parent 3 nontraditional students
4	5	5	100% female, 80% foreign-born: Grenada, Uzbekistan, St Lucia, Jamaica, USA	27–52	3 African-Caribbean 1 Caucasian 1 Black	Biology	3 seniors 2 juniors 4 nontraditional students 4 parents
5	7	7	14% male, 86% female, 14% foreign-born	22–33	4 Latin a/o 3 Black	Biology	2 sophomores 4 seniors 1 junior 1 nontraditional student 1 parent

Teacher-identity profiles of participants

Participants who completed the TIPS program thus far could be classified into three categories: those who shifted into teaching, those who had the most significant potential to shift, and those who did not pursue a teaching career (Table 1.3). We've learned that there are specific attributes and factors that influence the decision to teach. For many participants, finance, self-confidence, mentoring, and familial obligation remain consistent variables when they are deciding whether to become teachers.

Those participants who did not become teachers reported positive outcomes, crediting the experience with dispelling stereotypes about the teaching profession, allowing them to learn about the process of teaching, building their confidence to enable them to shift identities and see themselves as teachers, as well as providing classroom experiences with teachers and with students.

The participants with the greatest potential to shift were nontraditional students with responsibilities as breadwinners for their families. In some cases, balancing the expectations of family and finances proved challenging, and the tension in choosing between wanting to try teaching instead of a traditional high-paying job in healthcare led to the unresolved conflict to choose. One year after participating in the program, some participants still reported being undecided between teaching and other career options.

There were two common themes among those who became teachers; those participants were nontraditional students, ages 24–33, and motivated by initial interests and earlier exposure to teaching. Several participants worked as tutors before joining the TIPS program, suggesting its role as a reinforcing factor in their decision-making. Others worked in public school settings or with students but did not see themselves as teachers until they gained exposure to being in the classroom, working on lesson plans, and teaching a lesson to students. As co-directors of the TIPS program and college faculty, we also served as mentors and supported participants who showed potential to become teachers by providing information about graduate programs, teaching fellowships, and job opportunities with the hope of sparking a change.

In 2019, we selected four students to present with us at a research conference and receive additional mentoring; this was an invaluable experience for the mentees and mentors alike. While on the trip, we listened to our students in their varied Caribbean accents talking about their very full lives. We laughed as we reflected on which island we grew up on or visited as children. The stories resonated as all of the professors and students shared a common heritage and culture that was familiar. Most of our conversations during the trip to the conference were open-ended, explorative, and seemingly casual. We ate dinner and lunch together and spent most of our days either in conference sessions or sightseeing.

Jazmin, a biology major and a working mom, constantly balanced life, work, and school. For her, working was a necessity, not an option, as she cared for

Table 1.3 Profiles and attributes for the identity-shift of STEM majors

Identity shift from STEM major to STEM teacher	*Profile and attribute factors impacting decisions*	*Participants' comments*
Shifted into Teaching	Imani was a full-time student who immigrated from Trinidad and Tobago to New York; she worked as a part-time after-school tutor at a public elementary school. She was influenced by her family to become a nurse, a career tradition in her family, but she admitted she thought about a teaching career. As a high-achieving STEM student, she could pursue either career path, but after completing the TIPS program opted to teach. Imani was most influenced by the science teachers she observed in the classroom. Imani obtained her master's in teaching and has taught high school for over three years.	*"I enjoyed observing the class and student interactions, one thing that also I enjoyed was listening to the stories of Ms. Wave and another science teacher on how they went through the process of becoming teachers and what influenced them to choose the career path that they did. Like myself, teaching was not what neither of them thought that they would be doing. Both teachers were on the path to medical school when they decided to give teaching a try. By doing this, they both realized that they had a passion for this newfound craft and this shined through in both of their classrooms. I also enjoy learning about first-hand experiences with programs such as New York Teaching Fellows and Teach for America, which both women participated in order to become educators."*
	Jabari was a full-time student, full-time employee, spouse, and parent. He immigrated from Nigeria to the United States and was a biology major interested in environmental science. Before participating in the TIPS program, he wanted to become a medical doctor. The TIPS programs' professional development workshops resonated most with Jabari. After he participated in the program, he shifted to teaching and has worked as a biology/special education teacher for two years in a high school. He now aspires to become a high school principal.	*"The facilitators of the workshops are great because they were from diverse backgrounds and they came with interesting topics, interesting style of teaching, an incredible wealth of knowledge and an openness to help even during after-hours. They were true gurus in their respective fields. They are one of the reasons why I want to teach."*

Had the Greatest Potential to Shift	Jazmin was a biology major, mathematics minor, Math tutor, health aide, and mom. Her career goal was to become a nurse for several reasons; the heavy influence of her Jamaican family, of whom several people were nurses, believed it to be a well-paying job, the practicality of finding a position to take care of her family, and the lack of information about teaching as a career. Before participating in the TIPS program, Jazmin liked tutoring Math but felt conflicted because her family pressured her to pursue a high-salaried job such as nursing. Jazmin expressed that she wanted to become a Math teacher and was excited about the idea. During our follow-up a year later with Jazmin, she revealed that she worked as an assistant college laboratory technician to earn extra income, a position that utilizes her science background. The ability to take care of her family was essential to Jazmin, and she, therefore, prioritized financial stability. After participating in the TIPS program, Jazmin continued on the path of becoming a Nurse and did not become a teacher. She later applied to nursing programs but, to date has yet to become a nurse.	*"I never thought of teaching as a career seriously because . . . teachers don't make money,"*
	Nala is currently a senior biology major and a mother who works as a school safety agent. She was also heavily influenced to pursue nursing due to the occupation demand and lucrative salary. Before participating in the TIPS program, she aspired to be a Nurse. She was most excited to work with kids in the classroom during the program. One year after participating in the TIPS program, as Nala prepares to graduate with a Baccalaureate degree in biology, she is still torn between a career in nursing and teaching.	*"I think the most impactful part was gaining the experience to actually interact and work with the kids because I was really curious to know what it would feel like . . . would I be able to um see myself working in a classroom with the kids . . . would they listen to me?"*

(Continued)

Table 1.3 (Continued)

Identity shift from STEM major to STEM teacher	*Profile and attribute factors impacting decisions*	*Participants' comments*
Did not Pursue Teaching	Afiya was a biology major who planned to pursue a career in medicine, specializing in hematology. She participated in the TIPS program for two years and was very open to the experience. During her participation in the TIPS program, she benefited most from the professional development workshops. Two years after participating in the TIPS program, works as a healthcare analyst, but aspires to be a software engineer.	"*The TIPS program has been an amazing opportunity to connect and learn from professional educators. The workshops give insight into what makes specific educators great teachers. I learned a lot about teaching styles, methods, and ideologies that I was able to use and implement in my presentation at the end of the program.*" She added that "*Having now participated in my second year of the TIPS program, I can pinpoint my growth in the lesson plans that I created my first year versus the lesson plan I created during my second year. Granted, both lesson plans were geared toward two different age groups/grade levels, I was still able to recognize my growth through my end-of-the-program presentations. The TIPS program has allowed me to be more confident in front of a group of professional people and present a lesson plan that first started as a thought, idea, and concept and for that I am thankful.*"
	Evelyn participated in the TIPS program as a senior biology major who aspired to be a medical doctor. She was very confident as a STEM major who was sure of her career path but was interested in learning more about teaching. Evelyn was impressed by the teachers she observed in the classrooms and reflected on that experience as very impactful. Evelyn was excited about her classroom experience, and appreciated the real-world experiences that the teachers used to teach science. Two years after her participation in the TIPS program, Evelyn enrolled into a medical program with the goal of becoming a pediatrician.	"*Participating in the TIPs program has been very rewarding. Beginning this venture, outside of my comfort zone, was needed because it gave me perspective on what else can be a possibility, as a future career. It exposed me to students who enjoy learning and those who needed some help getting there. I was given an inside look at how a teacher preps for their class, how they try to implement the best learning styles for their students.*" "*Walking into . . . Ms. L's class, the first thing I saw was her, the principal and some of her students diligently working on multiple science projects. The students' eagerness to know what was going on, answering questions that were asked correctly, and why certain things are happening was refreshing. You can tell that they genuinely cared about the students and only want the best for them.*"

her young child. While we had career aspirations to teach, she was concerned about having the financial resources and time to take additional courses, or complete a master's degree, and teaching certificate. When we encouraged her to look into alternative certification routes, the uncertainty of a steady paycheck was too daunting and posed as a barrier to her taking the next step. It would take time to make the shift; her family depended on her financially, and this was her priority. The other three women on the trip were all over 21 years old, and while they were not mothers, they also had myriad obligations outside of school. Working was a regular part of most of their lives, and balancing work, school, and life was a common practice. TIPS participants are nontraditionally college students who attend a commuter college. They are challenged to make choices as working moms or working college students with family who often depend on them.

Imani accepted a job offer to teach science at a charter school. Her shift was intrinsic and organic, as we never force our students to take any particular path. When we asked her what caused her to shift from wanting a position in the medical field to becoming a teacher, she explicitly detailed the time she spent in one of our partner teacher's classrooms working with real students and the mentorship she received from faculty mentors. What the professors viewed as casual conversations were actually life-altering for her. Having the space to ask questions, and explore her ideas and aspirations without judgment, had given her the freedom to choose teaching. On her end-of-year reflection, Imani comments:

> *I enjoyed observing the class and student interactions, one thing that also I enjoyed was listening to the stories of Ms. Wave and another science teacher on . . . the process of becoming teachers and what influenced them to choose the career. . . . Like myself, teaching was not what neither of them thought that they would be doing. Both teachers were on the path to medical school when they decided to give teaching a try.*
>
> *Getting the opportunity to participate in [TIPS] has blessed me in more ways than I can imagine. Hearing the stories of STEM educators and gaining classroom experience has further ignited my passion for this field . . . . I am more than excited to continue participating. . . , and I look forward to continuing my journey towards becoming a science teacher.*

For students like Imani, TIPS was transformative. Being in a real classroom with dynamic STEM teachers allowed them to experience teaching in a real way. She began to understand that she might be able to use her STEM background outside of the medical field. Each year, students commented on the power of being in a real classroom and receiving mentorship from the program co-directors and the classroom teachers. We find that having the full scope of the TIPS experience has indeed opened our students to a new world of teaching.

Potential and possibilities

Another group of students we have encountered over the course of TIPS are those who have the potential and desire to teach but do not for various reasons. As we learned above, Jazmin's family expectations played an obstructive role in the decision to teach. As a working mother, she also had to provide for her child and felt as though any interruption of her income posed a serious threat to her family. This meant that additional schooling and pursuit of a new degree seemed too overwhelming at times. We were often puzzled by Jazmin's hesitation, as her passion to teach math was evident. After the pandemic, many STEM teaching positions became vacant because of the mass exodus caused by teachers retiring or leaving the field. We contacted Jazmin about employment opportunities, but she was nonresponsive. While we cannot say with certainty what deterred her, previous conversations about caring for her family coupled with family influence to pursue a career in medicine provide some insight into why her potential to teach remains unrealized.

Open, but staying the course

A third group of students that we find in project TIPS are those who are open to the idea of teaching, and would even consider it second career or part-time endeavor, but remain committed to the medical field. We emphasized to all students that we simply want to expose them to alternate career options; we do not mandate or force teaching, so it is always clear to students that we support their career choice. Students such as Evelyn and Afiya were more open to the idea of teaching but remained in the healthcare field after graduation. Upon graduating, Evelyn applied for and was accepted into medical school; she explained that she saw teaching as something she could do in the future. In her end-of-year reflection, she writes, "*I believe a program like this is necessary to give college students more exposure to their options and to not count teaching out because of the negative remarks. I'm happy I was able to be a part of the pilot program and looking forward to furthering my education in teaching.*" While she did not choose to teach, her sentiments about the type of exposure that TIPS offers provide insight into the benefits of such a program, even for those who become healthcare practitioners.

Nala—"I can do this and it's fun"

Nala is currently a senior biology major and a mother who works as a school safety agent. She was also heavily influenced to pursue nursing due to the occupation demand and lucrative salary. Before participating in the TIPS program, she aspired to be a nurse. She was most excited to work with kids in the classroom during the program. She reflected:

> *I think the most impactful part was gaining the experience to actually interact and work with the kids because I was really curious to know what it would feel like. Would I be able to um see myself working in a classroom with the kids? Would they listen to me?*

One year after participating in the TIPS program, as Nala prepares to graduate with a Baccalaureate degree in biology, she is still torn between a career in nursing and teaching. In her end-of-year reflection, she discusses the ways in which TIPS has increased her passion for teaching:

> *I thoroughly enjoyed everything that was parted to me by the instructors in this field of science, especially observing the youths and how they reacted to it. This is most certainly where I want to be. The beautiful plants that blossomed in the garden, as well as the passion that has grown inside of me, has surely put me on the right path of being an excellent biology teacher.*

A core aspect of the TIPS program is to help clarify what it means to be a teacher. We made this a vital element of the program because of the myths circulating about the teaching profession, specifically teaching in urban schools. The TIPS participants had strong and validated identities as STEM majors. Most had grade point averages above 3.2 and aspirations to pursue healthcare and medicine careers requiring high academic rigor and excellence. They were confident as STEM students yet lacked the confidence to teach STEM content. Part of reframing the myths about teaching is to help TIPS participants see themselves as STEM educators. The shifting professional identities of STEM majors is an essential factor when recruiting STEM majors of color (Lawrence et al., 2019). For example, across TIPS cohorts who responded to the survey, several noted apprehension and caution about teaching as a possible career option. Some common responses include:

- *I never saw myself in a position as a teacher*;
- although I do love teaching and tutoring, I just never envisioned myself ever becoming a teacher;
- I don't see me being a good teacher at all;
- the time it would take to complete the process of becoming a teacher;
- the requirement [*sic*] for teaching;
- there's [*sic*] so many exams you have to take to be a teacher and you have to always update your certifications throughout your years of teaching; and
- the concern of starting teaching a little late in life.

Teacher salary was consistently one of the top four decision-making factors influencing participants' decisions. On average, 37% of participants report that becoming a teacher is a financial decision. The qualitative responses below show that STEM majors of color weigh the long-term possibility of a sustainable lifestyle that can support them and their family.

- *My family would want me to get the job that makes the most money. Also [I] am concerned about opportunities for upward mobility in teaching. I really wouldn't want to just stay in the classroom for all my career.*
- I do not feel confident that I will be able to support the family I plan to have on a teachers [*sic*] salary.
- I feel as if teachers are not truly compensated for their time and effort they pour into children (especially with continuous budget cuts, and not taking into account the hours outside the classroom spent grading papers and creating new lessons).
- I think I'm nervous about creating a future in something I love, but not being financial [*sic*] equipped in living off of my salary.

References

Abe, E. N., & Chikoko, V. (2020). Exploring the factors that influence the career decision of STEM students at a university in South Africa. *International Journal of STEM Education, 7*(60). https://doi.org/10.1186/s40594-020-00256-x

Adamuti-Trache, M., & Andres, L. (2008). Embarking on and persisting in scientific fields of study: Social class, gender, and curriculum along the science pipeline. *International Journal of Science Education, 30*(12), 1557–1584.

Albert Shanker Institute. (2015). *The state of teacher diversity in American education.* Retrieved June 14, 2019, from www.shankerinstitute.org/sites/shanker/files/The%20State%20of%20Teacher%20Diversity_0.pdf

Ali-Hassan, H. (2009). *Social capital theory* (Chapter XXIV, pp. 420–433). IGI Global. http://biblio.uabcs.mx/html/libros/pdf/11/24.pdf

Archer, L., DeWitt, J., & Osborne, J. (2015). Is science for us? Black students' and parents' views of science and science careers. *Science Education, 99*(2), 199–237.

Archer, L., DeWitt, J., Osborne, J., Dillon, J., Willis, B., & Wong, B. (2012). Science aspirations and family habitus: How families shape children's engagement and identification with science. *American Educational Research Journal, 49*(5), 881–908.

Archer, L., DeWitt, J., & Wong, B. (2014). Spheres of influence: What shapes young people's aspirations at age 12/13 and what are the implications for education policy? *Journal of Education Policy, 29*(1), 58–85.

Aschbacher, P. R., Ing, M., & Tsai, S. M. (2014). Is science me? Exploring middle school students' STE-M career aspirations. *Journal of Science Education Technology, 23*, 735–743.

Aschbacher, P. R., Li, E., & Roth, E. J. (2010). Is science me? High school students' identities, participation and aspirations in science, engineering, and medicine. *Journal of Research Science Teaching, 47*, 564–582.

Atherton, G., Cymbir, E., Roberts, K., Page, L., & Remedios, R. (2009). *How young people formulate their views about the future*. Department for Children, Schools and Families.

Bardelli, E., & Ronfeldt, M. (2021). Workforce outcomes of program completers in high-needs endorsement areas. *American Journal of Education, 128*(1), 59–93. https://doi.org/10.1086/716486

Beaudoin, C. R., Johnston, P. C., Jones, L. B., & Waggett, R. J. (2013). University support of secondary STEM teachers through professional development. *Education, 133*(3), 330–339.

Bettinger, E., & Baker, R. (2011). *The effects of student coaching in college: An evaluation of a randomized experiment in student mentoring* (NBER Working Paper No. 16881). National Bureau of Economic Research.

Bourdieu, P., & Passeron, J. (1977). *Reproduction in education, society, and culture*. Sage Publishing.

Callender, C. (2020). Black male teachers, white education spaces: Troubling school practices of othering and surveillance. *British Educational Research Journal, 46*(5), 1081–1098. https://doi.org/10.1002/berj.3614

Casey, L., Di Carlo, M., Bond, B., & Quintero, E. (2015). *The state of teacher diversity in American education*. Albert Shanker Institute. https://www.shankerinstitute.org/resource/teacherdiversity

Cherng, H. S., & Halpin, P. F. (2016). The importance of minority teachers: Student perceptions of minority versus White teachers. *Educational Researcher, 45*(7), 407–420. https://doi.org/10.3102/0013189X16671718; https://journals.sagepub.com/doi/full/10.3102/0013189X16671718

Cherng, H. S., Halpin, P. F., & Rodriguez, L. A. (2022). Teaching bias? Relations between teaching quality and classroom demographic composition. *American Journal of Education, 128*(2), 171–201. https://doi.org/10.1086/717676; https://www.journals.uchicago.edu/doi/full/10.1086/717676?journalCode=aje

Chism, D. (2022). Building a diverse teacher pipeline starts with students. *Educational Leadership, 80*(2), 34–39.

Claussen, S., & Osborne, J. (2012) Bourdieu's notion of cultural capital and its implications for the science curriculum. *Science Education, 97*(1), 58–79.

Coleman, J. S. (1988). Social capital in the creation of human capital. *The American Journal of Sociology, 94*(Suppl), S95–S120. https://doi.org/10.1086/228943

Curran, F. C. (2017). Teach for America placement and teacher vacancies: Evidence from the Mississippi Delta. *Teachers College Record, 119*, 1–24.

Dabney, K., Chakraverty, D., & Tai, R. H. (2013). The association of family influence and initial interest in science. *Science Education, 97*(3), 395–409.

D'Amico, D., Pawlewicz, R. J., Earley, P. M., & McGeehan, A. P. (2017). Where are all the Black teachers? Discrimination in the teacher labor market. *Harvard Educational Review, 87*(1), 26–49.

Delpit, L. (2006). *Other people's children: Cultural conflict in the classroom* (1st ed.). The New Press.

Delpit, L., & Dowdy, J. K. (2008). *The skin that we speak: Thoughts on language and culture in the classroom*. The New Press.

Devari, S., Perkins-Hall, S., & Abeysekera, K. (2017). Tested strategies for recruiting and retention of STEM majors. *International Conference Educational Technologies*, 44–50.

DeWitt, J., Archer, L., & Osborne, J. (2013). Nerdy, brainy and normal: Children's and parents' constructions of those who are highly engaged with science. *Research in Science Education, 43*, 1455–476.

Dilwort, M. E., & Coleman, M. J. (2014). *Time for a change: Diversity in teaching revisited.* National Education Association. https://vtechworks.lib.vt.edu/bitstream/handle/10919/84025/ChangeDiversityTeaching.pdf?sequence=1&isAllowed=y

Donaldson, M. L., & Johnson, S. M. (2011). Teacher for America teachers: How long do they teach? Why do they leave? *Phi Delta Kappan, 93*(2), 47–51.

Douglas, E., & Khandaker, N. (2015). 8 promising practices for recruiting diverse educator talent. *Education Week, 34*(34), 26.

DuBois, D. L., Portillo, N., Rhodes, J. E., Silverthorn, N., & Valentine, J. C. (2011). How effective are mentoring programs for youth? A systematic assessment of the evidence. *Psychological Science in the Public Interest, 12*(2), 57–91. https://doi.org/10.1177/1529100611414806

Durlauf, S. N., & Fafchamps, M. (2005). Social capital. In P. Aghion & S. Durlauf (Eds.), *Handbook of economic growth* (Vol. 1, pp. 1639–1699). Elsevier.

Eccles, J. (2009). Who am I and what am I going to do with my life? Personal and collective identities as motivators of action. *Educational Psychologist, 44*(2), 78–89.

Evans, B. R., & Leonard, J. (2013). Recruiting and retaining Black teachers to work in urban schools. *SAGE Open, 3*(3), 1–12. https://doi.org/10.1177/2158244013502989

Ferrara, M., Talbot, R., Mason, H., Wee, B., Rorrer, R., Jacobson, M., & Gallagher, D. (2018). Enhancing undergraduate experiences with outreach in school STEM clubs. *Journal of College Science Teaching, 47*(6), 74–82.

Galaviz, S. (2020). *Designing STEM experiences for the family in order to develop STEM family habitus and capital* [Theses and Dissertations, Boise State University]. https://doi.org/10.18122/td/1686/boisestate

Gandhi-Lee, E., Skaza, H., Marti, E., Schrader, P., & Orgill, M. (2017). Faculty perceptions of student recruitment and retention in STEM fields. *European Journal of STEM Education, 2*(1), 1–11.

García, E., & Weiss, E. (2017). *Education inequalities at the school starting gate: Gaps, trends, and strategies to address them*. Economic Policy Institute.

Gay, G. (2018). *Culturally responsive teaching: Theory, research, and practice* (3rd ed.). Teachers College Press.

Gemici, S., Bednarz, A., Karmel, T., & Lim, P. (2014). *The factors affecting the educational and occupational aspirations of young Australians* (LSAY Research Report No. 66). National Centre for Vocational Education Research. www.ncver.edu.au/_data/assets/file/0021/9516/factorsaffecting-aspirations-2711.pdf

Gilmartin, S. K., Erika, L., & Aschbacher, P. (2006) The relationship between interest in physical science/engineering, science class experiences, and family contexts: Variations by gender and race/ethnicity among secondary students. *Journal of Women and Minorities in Science and Engineering, 12*(2–3), 179–207.

Gist, C. D. (2017). Voices of aspiring teachers of color: Unraveling the double bind in teacher education. *Urban Education, 52*(8), 927–956. https://doi.org/10.1177/0042085915623339

Green, S. L., & Jor'dan, J. R. (2018). Having our say: Building new identities through teacher preparation. *Multicultural Learning and Teaching, 13*(1), 1–4.

Green, S. L., & Martin, D. (2018). Playing the game: Recruiting Black males in teaching. *Multicultural Learning and Teaching, 13*(1), 1–14.

Green, T. L. (2017). Places of inequality, places of possibility: Mapping "opportunity in geography" across urban school-communities. *Urban Review*, *47*, 717–741. https://doi.org/10.1007/s11256-015-0331-z

Griffin, K. A., del Pilar, W., McIntosh, K., & Griffin, A. (2012). "Oh, of course I'm going to go to college": Understanding how habitus shapes the college choice process of Black immigrant students. *Journal of Diversity in Higher Education*, *5*(2), 96–111.

Griffin, K. A., Pérez, D., Holmes, A. P. E., & Mayo, C. E. P. (2010). Investing in the future: The importance of faculty mentoring in the development of students of color in STEM. *New Directions for Institutional Research*, *2010*(148), 95–103. https://doi-org.mec.ezproxy.cuny.edu/10.1002/ir.365

Grossman, P., & Loeb, S. (2010). Learning from multiple routes. *Educational Leadership*, 22–27.

Guha, R., Hyler, M. E., & Darling-Hammond, L. (2017). The teacher residency: A practical path to recruitment and retention. *The Education Digest*, *41*(1), 38–45.

Harackiewicz, J. M., Rozek, C. S., Hulleman, C. S., & Hyde, J. S. (2012). Helping parents to motivate adolescents in mathematics and science: An experimental test of a utility-value intervention. *Psychology Science*, *23*(8), 899–906.

Hill, D. A., & Gillette, M. D. (2005). Teachers for tomorrow in urban schools: Recruiting and supporting the pipeline. *Multicultural Perspectives*, *7*(3), 42–50.

Hill, N. E., & Tyson, D. F. (2009). Parental involvement in middle school: A meta-analytic assessment of the strategies that promote achievement in school. *Developmental Psychology*, *45*, 740–763.

Hubbard, K., Embry-Jenlink, K., & Beverly, L. (2015). A university approach to improving STEM teacher recruitment and retention. *Kappa Delta Pi Record*, *51*, 69–74. https://doi.org/10.1080/00228958.2015.1023139

Hussar, W. J., & Bailey, T. M. (2016). *Projections of education statistics to 2024* (NCES 2016 013). U.S. Department of Education.

Hutton, C. (2019). Using role models to increase diversity in STEM. *Technology and Engineering Teacher*, 16–19.

Hyde, J. S., Else-Quest, N., Alibali, M. W., Knuth, E., & Romberg, T. (2006). Mathematics in the home: Homework practices and mother-child interactions doing mathematics. *Journal of Mathematical Behavior*, *25*, 136–152.

Imazeki, J., & Goe, L. (2009). *The distribution of highly qualified, experienced teachers: Challenges and opportunities* (T. Q. Research and Policy Brief). National Comprehensive Center for Teacher Quality.

Jackson, C. K., & Bruegmann, E. (2009). Teaching students and teaching each other: The importance of peer learning for teachers. *American Economic Journal: Applied Economics*, *1*(4), 85–108.

Jones, G., Dana, T., LaFramenta, J., Adams, T. L., & Arnold, J. D. (2016). STEM TIPS: Supporting the beginning secondary STEM teacher. *Tech Trends*, *60*, 272–288. https://doi.org/10.1007/s11528-016-0052-5

Jones, R., & Cleaver, R. (2020). Recruiting and enrolling rural students: A model for increasing diversity in STEM. *Innovative Higher Education*, *45*, 253–263. https://doi.org/10.1007/s10755-020-09499-6

Joseph, G. (2016). Teach for America goes global. *The Nation*, 25–28.

The Journal of Blacks in Higher Education. (2019). *Georgia state program seeks to boost number of Black male teachers in STEM fields*. www.jbhe.com/2019/10/georgia-state-program-seeks-to-boost-number-of-black-male-teachers-in-stem-fields/

The Journal of Blacks in Higher Education. (2020). *15 HBCUs to have the opportunity to design STEM teacher preparation programs*. www.jbhe.com/2020/02/15-hbcus-to-have-the-opportunity-to-design-stem-teacher-preparation-programs/

Kinney, A. (2015). Compelling counternarratives to deficit discourses: An investigation into the funds of knowledge of culturally and linguistically diverse U.S. elementary students' households. *Qualitative Research in Education*, *4*(1), 1–25. https://doi.org/10.4471/qre.2015.54

Kraft, M. A., & Papay, J. P. (2014). Can professional environments in schools promote teacher development? Explaining heterogeneity in returns to teaching experience. *Educational Effectiveness and Policy Analysis*, *36*(4), 476–500.

Kunz, J., Hubbard, K., Beverly, L., Cloyd, M., & Bancroft, A. (2020). What motivates STEM students to try teacher recruiting programs? *Kappa Delta Pi Record*, *56*(4), 154–159. https://doi.org/10.1080/00228958.2020.1813507

Ladson-Billings, G. (2009). *The dream-keepers: Successful teachers of African American children* (2nd ed.). Jossey-Bass.

Lancee, B. (2012). Social capital theory. In B. Lancee (Ed.), *Immigrant performance in the labour market: Bonding and bridging social capital*. Amsterdam University Press.

Lawrence, S. A., Johnson, T., & Small, C. (2019). Watering our own lawn: Exploring the impact of a collaborative approach to recruiting African Caribbean STEM majors into teaching. *Journal of Negro Education*, *88*(3), 391–406. https://doi.org/10.7709/jnegroeducation.88.3.0391

Lee, A., Henderson, D. X., Corneille, M., Morton, T., Prince, K., Burnett, S., & Roberson, T. (2022). Lifting Black student voices to identify teaching practices that discourage and encourage STEM engagement: Why #Black teachers matter. *Urban Education*, 1–33. https://doi.org/10.1177/00420859211073898

Leggett-Robinson, P. M., Villa, B., & Davis, N. C. (2017, June). *Board #87: Native-born and foreign-born black students in STEM: Addressing STEM identity and belonging barriers and their effects on STEM retention and persistence at the two year college* [Paper presentation]. 2017 ASEE Annual Conference & Exposition, Columbus, OH. https://doi.org/10.18260/1-2-27945

Lewis, C. W., & Toldson, I. (2013). *Black male teachers: Diversifying the United States' teacher workforce*. Emerald Group Publishing Limited.

Madkins, T. C. (2011). The Black teacher shortage: A literature review of historical and contemporary trends. *The Journal of Negro Education*, *80*(3), 417–427.

Massey, D. S., Mooney, M., Torres, K. C., & Charles C. Z. (2007). Black immigrants and Black natives attending selective colleges and universities in the United States. *American Journal of Education*, *113*(2), 243–271.

Mau, W. (2003). Factors that influence persistence in science and engineering career aspirations. *Career Development Quarterly*, *51*(3), 234–243.

Milgrom-Elcott, T. (2016). Three ways to recruit and train more teachers. *Tech Directions*, 34–35.

Moritz, M., & Weiss, E. (2018). Opinion: The U.S. doesn't have enough STEM teachers to prepare students for our high-tech economy. 4 steps toward addressing that shortage. *The 74*. www.the74million.org/article/opinion-the-u-s-doesnt-have-enough-stem-teachers-to-prepare-students-for-our-high-tech-economy-4-steps-toward-addressing-that-shortage/

Morrell, P. D., & Salomone, S. (2017). Impact of a Robert Noyce scholarship on STEM teacher recruitment. *Journal of College Science Teaching*, *47*(2), 16–21.

Moyer-Packenham, P. S., Kitsantas, A., Bolyard, J. J., Huie, F., & Irby, N. (2009). Participation by STEM faculty in mathematics and science partnership activities for teachers. *Journal of STEM Education, 10*(3–4), 1–20.

Munoz, S. M., Heilig, J. V., & Del Real, M. (2019). Property functions of whiteness: Counter-narrative analysis of teach for America and their partnership with Black and Latinx fraternities and sororities. *New Directions for Student Services, 165*, 61–71. https://doi.org/10.1002/ss.20294

National Comprehensive Center for Teacher Quality. (n.d.). *Recruiting quality teachers in mathematics, science, and special education for urban and rural schools* (Report). https://gtlcenter.org/sites/default/files/docs/NCCTQRecruitQuality.pdf

National Science Board, National Science Foundation. (2020). *Science and engineering indicators 2020: The state of U.S. science and engineering* (NSB-2020-01). https://ncses.nsf.gov/pubs/nsb20201/

Neally, K. (2022). An analysis of the underrepresentation of minoritized groups in science, technology, engineering, and mathematics education. *School Science and Mathematics, 122*(5), 271–280. https://doi.org/10.1111/ssm.12542

Nugent, G., Barker, B., Welch, G., Grandgenett, N., Wu, C., & Nelson, C. (2015). A model of factors contributing to STEM learning and career orientation. *International Journal of Science Education, 37*(7), 1067–1088.

Ogbu, J. U. (1992). Adaptation to minority status and impact on school success. *Theory into Practice, XXXI*(4), 287–295.

Ogbu, J. U., & Simons, H. (1998). Voluntary and involuntary minorities: A cultural-ecological theory of school performance with some implications for education. *Anthropology & Education Quarterly, 29*(2), 155–188.

Parsons, S. (2013). Cerritos college employs internships to boost STEM teacher supply. *Community College Week, 9.*

Patton, L. D. (2009). My sister's keeper: A qualitative examination of mentoring experiences among African American women in graduate and professional schools. *Journal of Higher Education, 80*(5), 510–537.

Patton, L. D., & Harper, S. R. (2003). Mentoring relationships among African American women in graduate and professional schools. In M. F. Howard-Hamilton (Ed.), *Meeting the needs of African American women* (New Directions for Student Services No. 104). Jossey-Bass.

Perry, A. (2015). Teach for America shows it's learned a lesson about diversity: Now, what's next? *Education Digest*, 48–50.

Quinn, D. M. (2020). Experimental effects of "achievement gap" news reporting on viewers' racial stereotypes, inequality, explanations, and inequality prioritization. *Educational Researcher, 49*(7), 482–492. https://doi.org/10.3102/0013189X20932469

Rahman, T., Fox, M. A., Ikoma, S., & Gray, L. (2017). *Certification status and experience of U.S. public school teachers: Variations across student subgroups* (National Center for Education Statistics 2017-056). National Center for Education Statistics, Department of Education.

Rice, J. K. (2013). Learning from experience? Evidence on the impact and distribution of teacher experience and the implications for teacher policy. *Education Finance and Policy, 8*(3), 332–348.

Rich, M. (2015, April 11). *Where are the teachers of color?* www.nytimes.com/2015/04/12/sunday-review/where-are-the-teachers-of-color.html

Rodrigues, S., Jindal-Snape, D., & Snape, J. (2011). Factors that influence student pursuit of science careers: The role of gender, ethnicity, family and friends. *Science Education International, 22*(3), 266–273.

Rogers-Ard, R., Knaus, C. B., Epstein, K. K., & Mayfield, K. (2012). Racial diversity sounds nice; systems transformation? Not so much: Developing urban teachers of color. *Urban Education, 48*(3), 451–479. https://doi.org/10.1177/0042085912454441

Rollock, N., Gillborn, D., Vincent, C., & Ball, S. J. (2015). *The colour of class: The educational strategies of the Black middle classes*. Routledge Publishing.

Ronfeldt, M., Loeb, S., & Wyckoff, J. (2013). How teacher turnover harms student achievement. *American Educational Research Journal, 50*(1), 4–36. https://doi.org/10.3102/0002831212463813

Sack, J., Quander, J., Redl, T., & Leveille, N. (2016). The community of practice among mathematics and mathematics education faculty members. *Innovation in Higher Education, 41*(2), 167–182. https://doi.org/10.1007/s10755-015-09340-9

Samuelson, C. C., & Litzler, E. (2016). Community cultural wealth: An assets-based approach to persistence of engineering students of color. *Journal of Engineering Education, 105*(1), 93–117. https://doi.org/10.1002/jee.20110

Small, M. L. (2006). Neighborhood institutions as resource brokers: Childcare centers, interorganizational ties, and resource access among the poor. *Social Problems, 53*(2), 274–292.

Stanton-Salazar, R. D. (1997). A social capital framework for understanding the socialization of racial minority children. *Harvard Educational Review, 67*(1), 1–40.

Svoboda, R. C., Rozek, C. S., Hyde, J. S., Harackiewicz, J. M., & Destin, M. (2016). Understanding the relationship between parental education and STEM course-taking through identity-based and expectancy-value theories of motivation. *American Education Research Association Open, 2*(3), 1–13.

Teo, T. W., & Ke, K. J. (2014). Challenges in STEM teaching: Implications for preservice and inservice teacher education program. *Theory into Practice, 53*, 18–24. https://doi.org/10.1080/00405841.2014.862116

U.S. Department of Education, National Center for Education Statistics. (2019). *Digest of education statistics, 2018* (NCES 2020-009, Chapter 3). NCES.

Villegas, A. M., & Irvine, J. J. (2010). Diversifying the teaching force: An examination of major arguments. *Urban Review, 42*, 175–192.

Vilson, J. L. (2016). The need for more teachers of color. *The Education Digest*, 17–26.

Wallace, D. L., & Gagen, L. M. (2020). African American males' decisions to teach: Barriers, motivations, and supports necessary for completing a teacher preparation program. *Education and Urban Society, 52*(3), 415–432. https://doi.org/10.1177/0013124519846294

Wasserman, N. H., & Rossi, D. (2015). Mathematics and science teachers' use of and confidence in empirical reasoning: Implications for STEM teacher preparation. *School Science and Mathematics, 115*(1), 22–34.

Whang-Sayson, H., Daniel, J. C., & Russell, A. A. (2017). A serendipitous benefit of a teaching-exploration program at a large public university: Creating a STEM workforce that supports teachers and public education. *Journal of College Science Teaching, 47*(1), 24–30.

Will, M. (2018). N.C. district trains its own STEM teachers. *Education Week, 37*(18), 9–11.

Yosso, T. J. (2005). Whose culture has capital? A critical race theory discussion of community cultural wealth. *Race Ethnicity and Education, 8*(1), 69–91. https://doi.org/10.1080/1361332052000341006

2 Pathways into teaching

Considerations for advancing recruitment practices

Recruiting STEM majors into teaching begins with strategic and intentional activities designed to increase awareness of the profession (Hubbard et al., 2015) so that prospective teachers develop a connection that helps them identify with the profession. For example, job shadowing mentor teachers helps students "critically evaluate the profession and their own strengths and weaknesses relative to teaching" (p. 71). These kinds of field-based observations, where prospective teachers see first-hand what it's like to be teachers, help STEM majors to reframe their beliefs about teaching, receive real-time one-on-one mentoring from in-service teachers, and reflect on their own strengths, including content knowledge, cultural beliefs, and affinity toward teaching. Having a long-term mentor is more beneficial to developing stronger confidence than the short-term mentoring received in traditional teacher preparation programs (Hubbard et al., 2015).

We also believe that helping future teachers develop a 'teacher identity' is an important part of becoming a member of a community of practice (Coddington & Swanson, 2019). Teacher identity is a long, multi-step process that occurs over time and is influenced by prior experiences as students, specifically recollections of previous teachers, beliefs about teaching, self-perception about the ability to teach, experiences in teacher preparation programs, and the ability to self-assess and reflect (Chong et al., 2011; Dassa & Derose, 2017; Garza et al., 2016) and situate themselves within the school context (Coddington & Swanson, 2019). For STEM teacher candidates, negotiating their teaching identity was shaped through immersive field-based experiences observing classroom teachers to learn how to build rapport with students, designing and co-planning lessons with teachers, and working directly with students to help students with content (Coddington & Swanson, 2019). For some pre-service teachers, analysis of teaching videos raised some tensions between their beliefs about a professional image of a teacher, their lived experiences, and their attempts to enact their own teaching identity in relation to their perceptions of what it means to be a good teacher (Gallchoir et al., 2018).

DOI: 10.4324/9781003414452-3

Alternate route programs

There are different pathways into the teaching profession, with oversight by different state regulations (Grossman & Loeb, 2010). Alternate route programs "typically enable individuals with a bachelor's degree to begin teaching as the teacher of record before completing all the coursework required for full certification" (Grossman & Loeb, 2010, p. 22). Most of these individuals enter teaching in urban rather than affluent suburban districts (Grossman & Loeb, 2010). Spurred by the global need for more teachers, alternate route programs are also found in the Netherlands and Israel (Grossman & Loeb, 2010).

Most recruitment programs attempting to address the need for more STEM teachers have focused on identifying individuals who already know they want to teach, namely career changers or alternative route programs, and grow-your-own models (Douglas & Khandaker, 2015; Milgrom-Elcott, 2016). Although increasing awareness of the teaching profession is often part of those programs, few programs prioritize increasing students' initial interest in teaching as a recruitment strategy for STEM majors (Hubbard et al., 2015). This overlooks a potential pool of STEM majors who have not considered teaching as a career option or have no understanding of the profession to even consider it.

New Jersey was one of a handful of states to formalize the alternate route teacher certification option in the mid-1980s (Klagholz, 2001). The expectation was to "hire talented liberal arts graduates who did not study education in college and pay them as full-salaried teachers" (Klagholz, 2001, p. 33). This model has remained consistent over the past three decades, with some slight variations that focused on specific areas of need. For example, programs have focused on recruiting STEM teachers from industry careers, grow-your-own models by recruiting and training prospective teachers locally, and targeted recruitment of individuals from diverse backgrounds to increase representation in urban schools (DiMaria, 2013; Grossman & Loeb, 2010; Lawrence et al., 2019; LoCascio et al., 2016; Milgrom-Elcott, 2016).

Urban teacher residencies have become a prevailing model for new teacher recruitment in urban districts across the Unites States (Table 2.1).

There is a long-standing tradition of how to recruit and prepare teachers. Traditional teacher education programs tend to have a replicable formula that works for people who know they want to teach. However, for those STEM majors who are not sure that they want to become teachers, we try to inspire them throughout our program. One key factor at the College is that the sophomore year is when STEM majors who apply to the nursing program learn whether they are accepted. In contrast to a traditional teacher education program where students enroll because they know they want to teach, TIPS seeks out students who may not have considered teaching as a profession and potential student populations who may have been overlooked in the past as potential teacher candidates. Through TIPS we've learned that in order to recruit and motivate STEM majors who matriculate from high school to college to become

Table 2.1 Examples of alternate route programs

Program name	*State*	*Focus/Features*
Boston Teacher Residency Program (Madkins, 2011)	Massachusetts	Recruit and retain minorities and teachers for high-need subjects such as special education, science, math, residents spend 1 year in the classroom with mentor master teachers, and commit to teach for 3 years
CalTeach—University of California (Whang-Sayson et al., 2017)	California	Recruit STEM majors to explore teaching Service learning courses and field experiences that allow participants to see how STEM is taught in local K12 schools Mentored by practicing master teachers
New Teacher Project (Honawar, 2007)	National/Various States	Target mid-career professionals seeking career change Teachers receive urban and rural experiences through partnerships Work with districts to streamline hiring practices
Noyce Program (Morrell & Salomone, 2017)	National	Recruit STEM teachers
Rider University's Teach (Milgrom-Elcott, 2016)	New Jersey	Recruits and trains professionals to teach STEM
Teach for America (Donaldson & Johnson, 2011; Joseph, 2016)	International	Recruit teachers for shortage areas, particularly in high-need urban and rural schools
Teaching Fellows (Grossman & Loeb, 2010)	New York	High-need certification areas in math, special education, and science
Teacher TRAC Program—Cerritos College (Parsons, 2013)		Recruitment, development, and preparation STEM internships—early fieldwork with teacher mentors in special education, career and technical education with discipline specific internship, summer teaching assistantship, and afterschool discipline specific tutoring

teachers, we have to start when they are in their first or second year of college. We've found that TIPS provides the framework for a long-term recruitment strategy, where we introduce the TIPS program to STEM majors in their first and second year, and then once they join the program, typically in their junior or senior year, we provide experiences that help to cultivate their identity around STEM education.

Learning what it means to teach

Undergraduates who are able to explore teaching as a profession gain numerous benefits (Whang-Sayson et al., 2017). Internship-like experiences help enhance content and pedagogical knowledge (Parsons, 2013). Although field-based internships are the hallmark of traditional teacher education programs, they provide exponential benefits for STEM majors who do not know much about the field of education beyond their own schooling experience.

To help increase retention in STEM fields, job shadowing and mentoring can address the issue that "many students lack familial STEM role models that could encourage them to pursue and persist in a STEM major" (Gandhi-Lee et al., 2017, p. 6), and parents in STEM careers are likely to encourage their child's development through K-12 STEM education (Hutton, 2019). Service learning and field-based internship opportunities have been instrumental in changing the perceptions of prospective teachers (Whang-Sayson et al., 2017). Whang-Sayson et al. (2017) report that over 200 STEM majors participate in the CALTeach program, which is designed as a service-learning experience to explore teaching. Evaluations show

> that the culminating project of teaching an inquiry-based lesson during their field experience plays a large role in shaping their views of themselves as potential future educators. Thus, preliminary assessment suggests that CalTeach indeed provides opportunities for students to explore and shape career choices. (p. 26)

Volunteerism has been a pathway into teaching (Milgrom-Elcott, 2016) for some career changers and students without academic education background (Whang-Sayson et al., 2017).

Career exploration has been facilitated through outreach activities such as STEM clubs that allow middle, high school, and college students to interact and increase their interest in STEM (Ferrara et al., 2018). Although many of the undergraduate students participating in STEM clubs for middle and high school students may not consider teaching as a career, they "take responsibility for research, planning, and execution of activities" . . . faculty mentors and lead teachers (Ferrara et al., 2018, p. 75) to assist with their planning, identifying instructional resources, and using effective communication skills.

To learn about their attitudes toward and their beliefs about education, Whang-Sayson et al. (2017) surveyed current undergraduates and former participants in a program called CalTeach, which was designed to recruit STEM teachers, including alumni who did and did not enter teaching. CalTeach includes five service-learning courses that allow STEM majors to explore teaching careers and to see how STEM is taught in local schools (Whang-Sayson et al., 2017). With increasing enrollments up to over 200 students each year, participating STEM majors "have consistently indicated . . . that the culminating project of teaching an inquiry-based lesson during their field experience plays a large role in shaping their views of themselves as potential future educators" (Whang-Sayson et al., 2017, p. 26). Although the program helped to shape the career choices of some students, for those who did not enter teaching, the program helped increase their understanding of education-related issues and their likelihood to volunteer or participate in future school-based activities and fostered a greater appreciation for teachers (Whang-Sayson et al., 2017). Additionally, research shows volunteering in school-related activities has been an instrumental mechanism for recruiting teachers, particularly those transitioning to teaching from previous careers (Milgrom-Elcott, 2016).

Learning what it means to be a teacher means attending to misconceptions about the profession, which is an element that is often overlooked in teacher preparation, particularly for our targeted participants, who are not from a teacher education space. Through TIPS we sought to raise awareness for programs to focus some aspect of their program on career development, which is an overlooked area. Project TIPS is different from other teacher education programs, in that students are not education majors and may not even end up teaching. Instead, our aim is to provide concrete pedagogic experiences, professional development, mentorship, and meaningful learning experiences that introduce STEM majors to the field. Many of our participants have chosen to become teachers after being in our program, and we believe the intentional work with real students, STEM PDs, and mentorship either caused a shift in students' career goals or further ignited the desire to teach for those who have not had the tools to navigate becoming a teacher.

TIPS is an active recruitment model. We are actively recruiting and seeking out students and finding spaces where to look for students to help diversify teaching: classroom exploration, PD, collaborative research, monthly mentoring, and fieldwork with students.

What is the same and different from other programs? This is a brief introduction. We will talk in detail about the program in the next section. Compare TIPS to other programs at MEC or other teacher preparation programs. TIPS is unique because we are focusing on only recruiting STEM majors compared to other education programs that focus on students who know they want to enter teaching and major in education. (1) Compare the five years of MEC education program and how they become STEM teachers. Include MEC data on who is majoring in STEM. Right now it's elementary. There is a gap that TIPS is meeting because it is focusing on middle/high school. (2) Compare to the Barbados model, which focuses on graduate students. Emphasize the uniqueness of the recruitment model. Similar to the Barbados program, TIPS students do not yet know that they want to be teachers but may have an inclination. What is the recruitment model? Talk about the biology class and why we made the shift. (3) When looking for students, recruitment should start early. First, we thought juniors and seniors—that's too late. Then we shifted to freshmen/sophomores—and explaining this is ideal.

Several studies have highlighted the importance of introducing students of color to STEM professions well before college (Constantine et al., 2007; Handwerker et al., 2020; Moreno et al., 2016; Shillingford et al., 2017; Tang et al., 2008). Previous research shows that programs that use an integrated STEM curriculum help students develop content knowledge and skills that will prepare them for STEM fields through engaging school-based and afterschool programs. These programs suggest that school counselors are instrumental in introducing females and students of color to STEM careers (Constantine et al., 2007; Shillingford et al., 2017; Tang et al., 2008). One area often overlooked during these early introduction career awareness programs is STEM education.

We designed the program with some specific activities based on research on preparing teachers. TIPS explicitly allows participants to move from STEM majors to STEM educators through our collaboration with the School of Education and the School of Science, Health, and Technology. The School of Education is a traditional four-year teacher education program that involves taking foundational education courses, taking subject area methods courses, engaging in field-based work with small groups, one-on-one or whole-class teaching for at least 100 hours, and a culminating experience of completing a year-long student teaching practicum with a minimum of 150 hours. Project TIPS takes the elements of the traditional program and modifies it for STEM majors who are in the School of Science, Health, and Technology and may not have previously considered teaching. Participants complete at least 20 hours of service learning per semester in a real STEM classroom, work with teachers to learn how to plan, instruct and assess students, as well as receive at least ten sessions of professional development each spring. We meet monthly with students to learn about their particular needs, challenges, and any additional support they would like. Partaking in a program like TIPS gives students additional career options that they might not have previously considered. Evelyn from Cohort 1 commented in her end-of-year reflection, "*Participating in the TIPS program has been very rewarding. Beginning this venture, outside of my comfort zone, was needed because it gave me perspective on what else can be a possibility, as a future career*." For her and others, being out of their "comfort zone" helped them envision new career options. Other participants shared this idea of career development. Nadia, a student in Cohort 3, shares:

> *[Participating] in the TIPS program this year I was able to grow and expand on my outlook of career opportunities as a graduating biology major. My previous personal options for applying my degree to prospective careers were solely focused on the medical field until now. As a participant, I was introduced to the possibility of utilizing my degree for a career opportunity as an educator. With the information provided by Dr. Small, Dr. Johnson, and Dr. Lawrence, I now know the tools an educator needs and the pathway required to pursue a career within the educational field. I am now confident that if chosen I will become a valuable educator within any STEM-focused curriculum.*

References

Chong, S., Ling, L. E., & Chuan, G. K. (2011). Developing student teachers' professional identities: An exploratory study. *International Education Studies, 4*(1), 30–38.

Coddington, L. R., & Swanson, L. H. (2019). Exploring identity of prospective math and science teachers through reflections in early field contexts. *Journal of Teacher Education and Educators, 8*(3), 207–228.

Constantine, M. G., Kindaichi, M. M., & Miville, M. L. (2007). Factors influencing the educational and vocational transitions of Black and Latino high school students. *Professional School Counseling, 10*(3), 261–265. www.jstor.org/stable/42732518

Dassa, L., & Derose, D. S. (2017). Get in the teacher zone: A perception study of preservice teachers and their teacher identity. *Issues in Teacher Education, 26*(1), 101–113.

DiMaria, F. (2013). Producing a new breed of teacher. *Education Digest*, 48–53.

Donaldson, M. L., & Johnson, S. M. (2011). Teacher for America teachers: How long do they teach? Why do they leave? Phi Delta Kappan, 93(2), 47–51.

Douglas, E., & Khandaker, N. (2015). 8 promising practices for recruiting diverse educator talent. *Education Week, 34*(34), 26.

Ferrara, M., Talbot, R., Mason, H., Wee, B., Rorrer, R., Jacobson, M., & Gallagher, D. (2018). Enhancing undergraduate experiences with outreach in school STEM clubs. *Journal of College Science Teaching, 47*(6), 74–82.

Gallchoir, C. O., O'Flaherty, J., & Hinchion, C. (2018). Identity development: What I notice about myself as a teacher. *European Journal of Teacher Education, 41*(2), 138–156. https://doi.org/10.1080/02619768.2017.1416087

Gandhi-Lee, E., Skaza, H., Marti, E., Schrader, P., & Orgill, M. (2017). Faculty perceptions of student recruitment and retention in STEM fields. *European Journal of STEM Education, 2*(1), 1–11.

Garza, R., Werner, P., & Wendler, L. F. (2016). Transforming from student to professional: Preservice teachers' perceptions. *New Waves Educational Research & Development, 19*(2), 19–35.

Grossman, P., & Loeb, S. (2010). Learning from multiple routes. *Educational Leadership*, 22–27.

Honawar, V. (2007). New Teacher Project brings holistic style to urban districts. *Education Digest*, 27(1) 23–27. Retrieved from. https://tntp.org/assets/documents/EdWeek_TNTP_Holistic_Style_formatted.pdf

Hubbard, K., Embry-Jenlink, K., & Beverly, L. (2015). A university approach to improving STEM teacher recruitment and retention. *Kappa Delta Pi Record, 51*, 69–74. https://doi.org/10.1080/00228958.2015.1023139

Hutton, C. (2019). Using role models to increase diversity in STEM. *Technology and Engineering Teacher*, 16–19.

Joseph, G. (2016). Teach for America goes global. *The Nation*, 25–28.

Klagholz, L. (2001). State policy and effective alternative teacher certification. *Education Digest*, 33–36.

Lawrence, S. A., Johnson, T., & Small, C. (2019). Watering our own lawn: Exploring the impact of a collaborative approach to recruiting African Caribbean STEM majors into teaching. *Journal of Negro Education, 88*(3), 391–406. https://doi.org/10.7709/jnegroeducation.88.3.0391

Leggett-Robinson, P. M., Villa, B., & Davis, N. C. (2017, June). Board #87: Native-born and foreign-born black students in STEM: Addressing STEM identity and belonging barriers and their eff ects on STEM retention and persistence at the two year college

[Paper presentation]. 2017 ASEE Annual Conference & Exposition, Columbus, OH. https://doi.org/10.18260/1-2-27945

LoCascio, S. J., Smeaton, P. S., & Waters, F. H. (2016). How induction programs affection the induction of alternate route urban teachers to remain teaching. *Education and Urban Society*, *48*(2), 103–125.

Madkins, T. C. (2011). The Black teacher shortage: A literature review of historical and contemporary trends. *The Journal of Negro Education*, 80 (3), 417–427.

Massey, D. S., Mooney, M., Torres, K. C., & Charles C. Z. (2007). Black immigrants and Black natives attending selective colleges and universities in the United States. *American Journal of Education*, 113 (2), 243–271.

Milgrom-Elcott, T. (2016). Three ways to recruit and train more teachers. *Tech Directions*, 34–35.

Moreno, N. P., Tharp, B. Z., Vogt, G., Newell, A. D., & Burnett, C. A. (2016). Preparing students for middle school through after-school STEM activities. *Journal of Science Education and Technology*, *25*(6), 889–898. www.jstor.org/stable/45151294

Morrell, P. D., & Salomone, S. (2017). Impact of a Robert Noyce scholarship on STEM teacher recruitment. Journal of College Science Teaching , 47 (2), 16–21.

Parsons, S. (2013). Cerritos college employs internships to boost STEM teacher supply. *Community College Week*, *25*(12), 9.

Shillingford, M. A., Oh, S., & Finnell, L. R. (2017). Promoting STEM career development among students and parents of color: Are school counselors leading the charge? *Professional School Counseling*, *21*(1b), 1–11. www.jstor.org/stable/90023616

Tang, M., Pan, W., & Newmeyer, M. D. (2008). Factors influencing high school students' career aspirations. *Professional School Counseling*, *11*(5), 285–295. www.jstor.org/stable/42732837

Whang-Sayson, H., Daniel, J. C., & Russell, A. A. (2017). A serendipitous benefit of a teaching-exploration program at a large public university: Creating a STEM workforce that supports teachers and public education. *Journal of College Science Teaching*, *47*(1), 24–30.

3 The TIPS program

A new pathway into teaching

Prioritizing recruitment

Our primary focus for the TIPS program is recruitment. With an overarching goal of recruiting STEM teachers of color to teach in urban schools, the program offers STEM majors mentoring, professional development, field experiences, and advisement. All program activities are guided by three objectives:

(1) To use targeted recruitment strategies for identifying, selecting, and engaging with STEM majors, including nontraditional students
(2) To mitigate the social, academic, and financial barriers that influence the decision-making of STEM majors of color when considering teaching as a profession
(3) To increase STEM majors' awareness and interest in teaching through professional development experiences

To achieve these goals, we work toward minimizing the silos between teacher education programs and STEM departments.

The program addresses the academic, financial, and social needs of students of color who are interested in pursuing careers in teaching by providing innovative academic and cultural enrichment opportunities to a cohort of STEM majors each year. Through these experiences, participants are exposed to pedagogical strategies that use inquiry-based, real-world math and science activities to build their pedagogical content knowledge and gain the required skills to pursue their initial licensure, which leads to a teaching career. The program also gives them access to mentoring opportunities as they interact with a wide range of educators and STEM professionals.

TIPS builds upon proven strategies and practices that have been successful in preparing STEM majors from underrepresented groups to be teachers. The program uses a year-long model to provide participants with tailored experiences designed to meet the needs of the targeted group. When recruiting, we typically identify low-income students of color who are STEM majors and have expressed interest in pursuing a teaching career. We also provide STEM majors

DOI: 10.4324/9781003414452-4

with an immersive experience to enhance their understanding of innovative pedagogical strategies to bridge the gap between STEM learning and STEM teaching. Through various mentoring and professional development opportunities, participants work with STEM professionals and near-peer mentors from diverse backgrounds who have similar professional interests to them.

Targeted recruitment

We begin the recruitment process during the sophomore year in a biology course on genetics. Targeting sophomores was a process that unfolded through convenience, trial, and error. We selected this course for targeted and aggressive recruitment efforts (10.1177/0013124519846294Wallace & Gagen, 2020) because it is taught by one of the program's co-directors, who at the time was also an adviser for biology majors. This provided convenient access to a potential pool where teacher educators could visit the class and talk about teaching as a career option. We also targeted this course because sophomore year is typically when biology majors at the college who are pursuing medical professions learn whether they are accepted into the nursing program. With an average of 1,100 biology majors annually and approximately 50 openings for the nursing program, the selection process for the nursing program is competitive and rigorous. Those not accepted for the nursing program learn about the results of their application to the program as sophomores and have to pivot as they consider alternative career options. Although we have also recruited juniors and seniors into the program, we learned that for many of them making the shift to think about STEM education at that point in their academic career is more challenging. Later, we'll discuss how we adjust the content to help graduating seniors think about graduate programs and alternate route options for their certification.

During the first three weeks of the semester, we used several methods to increase awareness of the program in our attempts to recruit a cohort of approximately ten participants. In addition to the targeted audience in the biology class, we also use recruitment strategies such as visiting classes, circulating and posting flyers in strategic places around the campus, announcements in the weekly college newsletter, hosting info sessions, and email blasts to STEM faculty.

The TIPS program is voluntary for high-performing undergraduate STEM majors. After participating in the year-long program, some decided to explore teaching as a career option. Over the past three years, 24% of participants self-identify as Black, 18% identify themselves as Black and Jamaican, 6% as African/Black Moroccan, 6% as Black/Nigerian, or 6% as Black/Trinidadian. Each year, the initial applicant pool includes mostly female students and a few male students; to date, only one male student has completed the program.

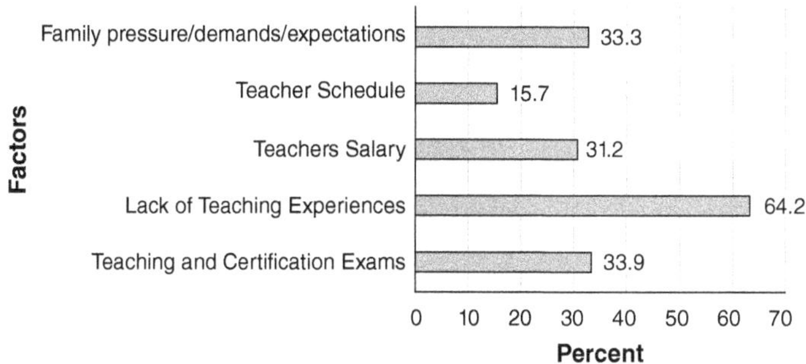

Figure 3.1 Factors influencing the decision to choose a career in teaching

When STEM majors start the program, we administer an initial survey (Appendix A) to gauge their interest in teaching, learn more about their career goals, gain insight on some of the factors that might be impacting decisions to enter teaching, and compare their perspectives to an end-of-year survey (Appendix B). Figure 3.1 shows that across cohorts, on the initial survey, lack of teaching experience was the most influential factor in their decision-making.

Some of the qualitative comments provided on the surveys helped shed some light on the tensions facing STEM majors of color, which seem to frame their decision-making. When asked to "provide more information about the factors you selected" on the survey, respondents from Cohort 2 said:

- *"I feel as if teachers are not truly compensated for their time and effort they pour into children (especially with continuous budget cuts, and not taking into account the hours outside the classroom spent grading papers and creating new lessons)."*
- *"Need hands on experience."*
- *"Although I do love teaching and tutoring, I just never envisioned myself ever becoming a teacher."*
- *"Although I want to be a doctor, my mom (along with other family members) pushed me into the field saying how great I'd be as a doctor. Teaching has never been an option for me till now."*
- *"In the beginning of School, I wanted to be a nurse I will love to get in the teaching field. I understand the importance of molding young minds."*
- *"I really want to pursue teaching. However when I started college that was not the plan. So "I need a plan how to transition from what I'm currently doing to teachers' education."*

- "*I also never saw myself in a position such as a teacher.*"
- "*My family would want me to get the job that makes the most money. Also am concerned about opportunities for upward mobility in teaching. I really wouldn't want to just stay in the classroom for all my career.*"
- "*I do not feel confident that I will be able to support the family I plan to have on a teacher's salary.*"
- "*The concern of starting teaching a little late in life.*"
- "*I tutor 1 ore 2 students in math and I always thought I never explain it the best way I could.*"

These comments echoed consistent themes that we've seen over the years. At the forefront of their decision-making are their confidence to teach their content to students, finances and supporting a family on a teacher's salary, their knowledge about the field of teaching, including opportunities for career advancement, and the familial pressures about a career in medicine. Over the years, we have used these comments to tailor orientation materials and topics shared when targeting sophomores.

Multidimensional mentoring

Purposeful advisement

As mentioned above, our recruitment efforts focus on STEM majors who are not accepted into the nursing program during the sophomore year. STEM faculty have shared that during informal conversations with advisees, biology majors who are not accepted into the nursing program graduate with biology degrees and may take one to two gap years or consider jobs well outside the STEM fields, such as retail or other industries. Anecdotal evidence from advisement sessions also indicates that many biology majors do not have a solid career path selected by sophomore year. They often default to majoring in biology with the hope of using the degree to pursue some form of healthcare. One TIPS participant from Cohort 3 said:

> *I decided to join the TIPS program at a time where I was very unsure of what I wanted to do with my life. Ever since I was a child, I always said I wanted to become a doctor, but as the years progressed my enthusiasm started to digress. The change in energy all started in 2019. That was the year where I completely lost a sense of who I was. I was sad and frustrated because I felt like I had already put so many years into my education, and somehow those years just seemed useless because I no longer knew what I wanted to do. Not only was I stressed, but I was no longer motivated to do anything.*
>
> *When I heard about the TIPS program, I immediately jumped on the opportunity. What drew me to the program was the fact that it was a*

> *program that would allow me to discover a different career path and a different aspect of STEM. Before entering into this program, I never thought about becoming a teacher as it has never interested me. However, since I did not know if I still wanted to go to medical school, I figured this could be a potential alternative.*

Students often feel discouraged after they have graduated and left the support of the college community and peers. With the absence of mentoring, many capable students do not pursue graduate school or medical school simply because it feels unattainable the longer they are away from the rigors of academia. Under the pressure of being the primary financial provider for the family, many students take available jobs that do not utilize their STEM degree or knowledge. We have found that when students in TIPS receive continued mentoring and faculty support after graduation, it often helps to guide and support them to apply to graduate programs and can create networking opportunities to find employment using their STEM degrees.

The advisement process starts with information sharing. We first identify STEM faculty who can share information with students in their courses. Then, the education faculty visit the class to meet with students to share information about teaching as a career. The presentation includes the education faculty's personal experiences transitioning from STEM to education, addressing students' concerns about making that shift, and identifying key processes to consider when becoming a teacher. Next, we schedule an information session for those who express interest in learning more about the TIPS program commitments and activities.

The program attracts students who at minimum have a curiosity about teaching. Though aware of the challenges of being a STEM major, we've learned from our surveys that respondents admitted to facing challenges as STEM majors but were not dissuaded from pursuing science and felt confident about choosing a STEM major despite the challenges. At the beginning of Year 3, three of four participants planned on pursuing careers in medicine and 25% in allied health. One respondent said, "*I have a passion for helping people, especially those with sickle cell.*" Another respondent said, "*I love helping people and want to become a doctor no matter how hard the courses are.*" Over the years, several program participants have detailed a greater purpose that was to be medically helpful to the family. Many of them reflect on how to transfer their passion for helping others from medicine to education.

By the end of a full year, participants either want to become teachers or are open to the possibility of becoming a STEM educator and pursuing options in graduate school. In alignment with previous research, we've also seen that some key factors that contribute to this shift include, but are not limited to, (1) exposure to real teaching practice and classrooms, (2) STEM faculty to introduce them to the program (3) varied PD emphasizing pedagogy, and (4) mentorship by in-service teachers and faculty. To be most impactful when working with our participants, we focused on structuring the program to provide hands-on exposure to teachers in classrooms, which students often rank as a crucial element of our program.

Each year, the TIPS participants are consistently a group of high-achieving STEM majors who were confident in their ability to pursue STEM careers but less so in teaching as a career. Many of the participants in the TIPS program lacked knowledge of teaching or had not considered a teaching career in part due to a lack of exposure and information. Over the year-long experience, they develop more confidence in their ability to teach or share their content knowledge by explaining concepts to others.

STEM majors in their junior year of undergraduate studies are invited to attend an orientation for the TIPS program. During the meeting, they are asked to complete an application if they are interested in participating. Individuals who apply and are selected to participate receive a stipend. Although several STEM majors typically express interest in the TIPS program and attend the orientation to learn more about the program, most participants have also expressed concerns about teaching and explained that they want more information specifically about the teaching profession. Every year, the consistent theme across all cohorts is that they did not previously consider teaching as a career option.

Mentoring

All the TIPS participants recruited for the program have been on a path toward a career in medicine and are at a crossroads. Although unsure of a career in medicine, they did not know how to proceed or identify alternatives. Charlene (all names are pseudonyms) said, "Ever since I was a child, I always said I wanted to become a doctor, but as the years progressed my enthusiasm started to digress . . . I was sad and frustrated because I felt like I had already put so many years into my education." A survey participant noted, "I really want to pursue teaching, however, when I started college that was not the plan. So I need a plan on how to transition from what I'm currently doing to teacher's education."

Some indicated that they lost a sense of purpose or sense of self. Anthony said, participating in the program was an "opportunity to discover a new career path." Christina said, "I believe a program like this is necessary to give college students more exposure to their options and to not count teaching out because of the negative remarks." She added, "Beginning this venture, outside of my comfort zone, was needed because it gave me perspective on what else can be a possibility, as a future career."

Throughout the academic year, the participants receive mentoring from faculty, in-service STEM teachers, peers, and STEM experts in the field. Mentoring included opportunities to interact and collaborate with peers on projects such as lesson planning and presentations, observe classroom instruction in a middle school science class and speak with a cooperating teacher, and learn from higher education faculty about how to transition from being a STEM major into teaching.

The year-long TIPS program consists of three core mentoring experiences:

- Monthly meetings with education and STEM faculty
- Mentor teacher (cooperating teacher) supporting fieldwork in a local school to observe classroom practice
- Professional development workshops in inquiry education led by STEM experts

Monthly Meetings. The monthly meetings have a fluid agenda that is adapted based on the needs of the group. At times, they are reflective of a support group meeting, professional development workshop, group advisement session, or a class with a mini lesson. Meeting topics have included lesson planning, questioning strategies using Bloom's Taxonomy, choosing and applying for a master's program, expectations for teacher certification exams, understanding teacher biases and curriculum, analyzing classroom teaching practices, educational histories and family dynamics. During monthly meetings, participants may read educational articles about various teaching styles and then discuss which approach is best suited for students in urban settings. They also watch videos of numerous teachers and compare teaching styles. The meetings also allow space for participants to reflect, ask questions, and share informal feedback on program experiences. For instance, when participants asked questions about how to engage different types of students and how to select and apply for master's programs, we focused specific meetings on these topics and shared materials and resources to assist them.

Given that STEM majors who enter teaching have little professional preparation in pedagogy, we focused on providing more exposure to the field of teaching and how they can transition into the profession. Faculty, classroom teachers, and peers have been invited to share their stories. For example, a former student who started college in the nursing program but transitioned to education spoke to participants about her experiences and her new teaching position. She provided insight about how she navigated the shift academically and personally without family support by having to work multiple part-time jobs to support herself financially. She said that as a nursing student, her "heart wasn't in it," and she was majoring in nursing "to make my mother happy." When she became a 4th grade teacher, she also talked about the rewards of inspiring young kids, the benefits of doing something she loves, and the job security of a full-time teaching position. A former TIPS participant was also invited back to share her experiences in the program and how it helped her transition to being a middle school science teacher.

We asked participants about the factors that they initially reported were influences on their decision to become a teacher. Initially, family was a significant factor in their decision to enter the healthcare field, 85.8% ($N = 6$). By the end of the year-long program, when asked if their family would respect their decision to become a teacher, 71.4% ($N = 5$) of participants agreed or strongly agreed that they would. Participants noted that correcting misconceptions among family members, such as a teacher's salary being significantly lower than a nurse's salary, helped when they tried to gain family members' respect for selecting a career in teaching.

Mentor Teachers. TIPS participants can work with current STEM teachers by observing in real classrooms, collaborating with them to develop lesson plans, and participating in workshops to support their understanding of teaching. Initial surveys showed that most TIPS participants expressed concerns about classroom management or specific concerns about teaching. During monthly meetings, they reflected on their own high school experiences as negative spaces with disruptive students. They also expressed trepidation about adequately explaining content to students. Some examples from the surveys include

- being able to communicate with children effectively, teaching them advance lessons without over teaching;
- how to properly relay information to students; and
- being able to work with students before seeing if this is the right area to venture into.

Although most TIPS participants focused on their lack of pedagogical knowledge or their efficacy teaching in their specific content area, some comments also reflected preconceived notions and biases about students in urban classrooms. After the classroom visits, participants in Cohort 1 were asked to visit a school and observe practices around classroom management, curriculum planning, and instructional strategies. Some participants noted the significance of relationships and how those relationships impacted the classroom environment. Robin said:

> *Her students were always engaged and were eager to participate in classroom activities. Seeing how excited the children were to see their teacher filled me with excitement as well. This taught me the importance of student/teacher relationships and how the profession of teaching is not primarily focused on academic, but on how well you relate to and get through to your students*. Even outside of her teaching period, students still flocked to her classroom.

As noted, most participants expressed trepidation about their efficacy in content knowledge and teaching content in school. They wanted field work that would allow them to "learn hands on," and they wanted more information about "how to get into programs that can help us become a teacher." The field work helped to increase the participants' perceptions about their ability to teach.

In end-of-year interviews, participants revealed that having experience in the classroom was among the most powerful and beneficial experiences. The experience helped dispel myths and preconceived notions about working with urban students.

> *Walking into her class, the first thing I saw was herself, the principal and some of her students diligently working on multiple science projects. The students' eagerness to know what was going on, answering questions that were asked correctly, and why certain things are happening was refreshing. You can tell that they genuinely cared about the students and only want the best for them.*
>
> (Afiya, Cohort 1)

For students who already had some background working with students or already knew they wanted to teach, the experience confirmed their decisions. Midway through the program, Lily said:

> *I am going to start teaching in the fall, so I felt like a program like this will either help me connect with people or get other development type stuff that I would need for myself for starting to work, so if I could get as much as information that I can as possible, would just go ahead and call it a day. I feel like for me, it helped me, personally it helped me.*

TIPS participants expressed apprehension about their knowledge and commented on the teacher's ability to teach the content. Although the participants were high-performing science majors and seniors in the biology program, many were concerned about the ability to teach the topic.

Daisy said the cooperating teacher "*knew the topic that was being taught like the back of her hand, and she was able to efficiently relay that information to her students.*" Another student, Robin said, "*Science is a very broad area and attempting to teach it is fairly difficult. It requires an in-depth understanding of the subject and the ability to teach it more than one way.*" She noted the importance of teacher planning to "*implement the best learning styles for their students and how to raise successful students.*" Robin, like other TIPS participants, gained pedagogical knowledge through field observation, which filled gaps in understanding around teaching and learning. In her reflection, Robin said:

> *Ms. Wave was teaching what "field of view in a microscope" was, and one student was still unclear. So she proceeded to give real world examples, then multiple practice questions until everyone understood.*

Cohort 3 developed lessons they implemented during virtual and in-person workshops with kids. One participant from Cohort 3 said:

> *I began this program with an overwhelming concern that I may not be able to achieve the end goal of creating a lesson plan and essentially teaching a chosen topic to all participants. However, with our mentors . . . we were both able to execute and deliver on the desired task. [The mentor] was very helpful with all of her "tips" and examples for producing a lesson plan and presentation [and] provided great suggestions and insight for different tools to use to keep your students involved when teaching a topic. With the guidance of our mentors and program leaders this resulted in our ability to produce our desired lesson plan and at the end we were both proud and confident with our growth.*

Working with real students

All participants were required to complete service learning with a highly effective science teacher in a partner middle school. The middle school selected has partnered with Malcom College's SOE and School of Science, Health, and Technology for several years and has a Science, Technology, Engineering, Arts, and Math (STEAM) focus; hence, this middle school was selected because of the long-term potential of this partnership. The demographics of the partnering school reveals a population of students of color, with 81% Black students, 15% Latinx, and 3% Asian. The school also included diversity in the areas of language and exceptionalities, with a population of 7% English Language Learners (ELL) and 25% students with special needs. Over 82% of students receive free and reduced lunch, an indicator of lower socioeconomics across the school district. Collaborative activities with local, urban STEM teachers at the school occurred simultaneously with virtual professional development workshops led by STEM experts.

Students in the first two years of the program visited a middle school science classroom to observe a STEM teacher in the sixth grade and, in Year 3, worked one-on-one remotely with a teacher to practice developing and teaching a lesson plan of their own to program participants.

In her reflection regarding the in-classroom experience, one participant stated:

> *Walking into [Ms. Wave's class], I saw her students diligently working on multiple science projects. The students' eagerness to know what was going on, answering questions . . . correctly. . . . You can tell that they genuinely cared about the students. . . . I even had a teaching moment where a student didn't know there was an exponent function on the calculator rather than to continuously multiply the same number.*

Some TIPS participants had experience working with children as babysitters and assistants in afterschool programs. However, although they had prior experience working with children, Daisy had some anxiety about completing the fieldwork. Participants were not sure about expectations for teachers and how teachers engaged students in science classrooms. Daisy "*enjoyed observing the class and student interactions*." She also commented on the teacher–student relationships. Although from her first visit she perceived the cooperating teacher to be "*a force to be reckoned with in her classroom*," she wrote in her reflection:

> *[S]tudents were always engaged and were eager to participate in classroom activities. . . . [S]eeing how excited the children were to see their teacher filled me with excitement as well. This taught me the importance of student/teacher relationships and how the profession of teaching is not primarily focused on academics, but on how well you relate to and get through to your students. Even when it was outside of her teaching period, students still flocked to her classroom and engaged in study hall and work* [sic] *on their science projects for the upcoming science fair.*

Despite her prior experience with children in an afterschool program, Daisy had a shift in her knowledge base as she observed a STEM teacher. She believed that the opportunity to observe the classroom helped alleviate some fears about teaching in urban classrooms. These fears consisted of knowledge to teach the subject and the ability to work with the students.

Professional development by STEM experts

Students in the TIPS program engage in a wide range of program activities, including monthly structured mentoring meetings, annual professional development workshops by STEM professionals on pedagogical approaches for teaching STEM content, and observing and working collaboratively with teachers in urban STEM classrooms. In TIPS, prospective STEM teachers had opportunities to work in the field. These opportunities made concrete and demystified what it means to be a classroom teacher. Here we explore what kinds of support Black STEM majors need to become classroom teachers and the kinds of experiences that shape the decision-making for STEM majors when determining whether to pursue teaching.

TIPS participants are STEM majors who do not receive a teaching degree or certification from the college. What we provide is an introduction to pedagogy and real-world teaching experiences.

Although the participants were STEM majors, few had teaching experience. Therefore, professional development focused on pedagogy. STEM faculty and practitioners are recruited to work with TIPS participants. Workshops facilitated by partnering STEM professionals emphasized strategies for teaching STEM subjects and developing and implementing inquiry-based lessons in the discipline.

Professional development workshops were offered by local STEM cultural institutions partnering on the project. For example, Brooklyn Botanical Gardens designed and delivered virtual workshops on botany, earth science, and environmental science. Participants were introduced to instructional strategies and curriculum resources for teaching STEM in middle and high school classrooms. These workshops along with the opportunity to work with mentor teachers created a multi-tiered professional development for the participants.

TIPS participants are predominantly biology majors; a few have minors in environmental science or math. During Year 3, more STEM faculty and teachers across disciplines were recruited to broaden participants' exposure to other subject areas, including technology/computer science, engineering, chemistry, mathematics, and earth science. For example, a male math faculty member was invited to join the program and offer summer workshops to Cohort 3. Inviting a male faculty member to participate aligned with Year 3 goals to place more emphasis on recruiting males of color, as "a positive role model in the form of a male educator of color has the potential to nurture the development of a positive self-image for males of color and turn the tide of academic achievement for this demographic" (10.1177/0013124519846294Wallace & Gagen, 2020, p. 416). During the first two years of the TIPS program, only two male students attended the information session in Year 1, and there have been no male participants. We disseminated lots of flyers (Figure 3.2) and held several info sessions to increase participation.

The professional development experiences provided by STEM field experts has helped to increase participants' pedagogical content knowledge and understanding how to teach inquiry-based lessons. For example, Cohort 3 participated in a virtual professional development program where five middle and high school classroom teachers were recruited and paired with two to three TIPS participants to design an integrated inquiry lesson. A workshop series on honeybees (Appendix C) included working as a team; they attended monthly workshops and subsequently thought about how to design lessons for students. One participant from Cohort 3 shared a reflection on the workshop. "*The time spent in the TIPS program was both nurturing and exciting. The workshops were worth the time spent and the growth possibility for a student in the TIPS program is endless because we have knowledgeable people to help and give us feedback.*" Another student from the same cohort said:

> *During my participation in the TIPS program this year I was able to grow and expand on my outlook of career opportunities as a graduating biology major student. My previous personal options for applying my degree to prospective careers were solely focused on the medical field until now. As*

a participant, I was introduced to the possibility of utilizing my degree for a career opportunity as an educator. With the information . . . , I now know the tools an educator needs and the pathway required to pursue a career within the educational field. I am now confident that if chosen I will become a valuable educator within any STEM-focused curriculum.

The learning experience that I had with the workshops taught me valuable information from both sides of the classroom. Having the point of view as a learning student acquiring knowledge from our PD instructors and also the brand new perspective on what being an educator entails is essential to being a successful teacher. I learned about the positive effects that origami has and also that to teach origami virtually to an inexperienced group takes a lot of patience, individual attention and expertise. . . . Witnessing all of our PD presenters taught me that there is an immense level of planning, preparedness, and knowledge of desired topic that is required for your targeted learning group.

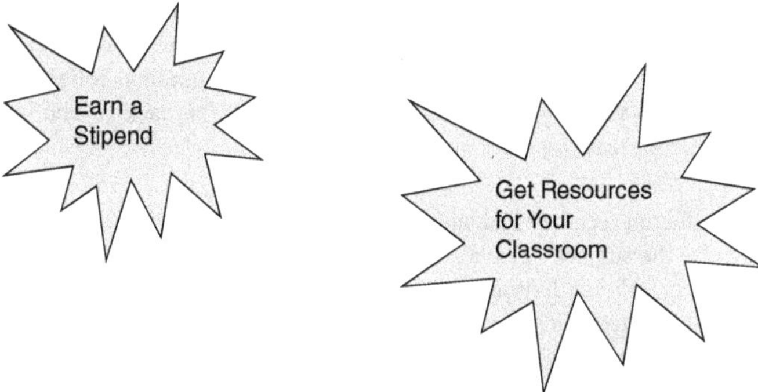

Are you interested in completing an action research project in your STEM classroom?

Join the TIPS Professional Leaming Community!
February - May 2021

1. Attend 2 monthly professional development workshops
2. Design and implement 1 project-based STEM unit in your classroom
3. Maintain an ePortfolio with materials and artifacts from the project
4. Receive a stipend for classroom supplies
5. Learn about resources for teaching STEM through active engagement
6. Mentor a preservice STEM teacher

This program is limited to 6 teachers.

Applications are due February 5, 2021.

Figure 3.2 Recruitment flyer for STEM faculty and teachers

Despite the benefits, participants required more scaffolding and guidance to make interdisciplinary connections to their content areas. When paired with Mary, who had prior experience as a teacher's aide, Charlene, who had no previous teaching experience, was overwhelmed when working on her lesson planning project. Charlene said:

> *There was a point where I was ready to quit and just get out of the program. I had no idea what I wanted to present and not knowing how to present a lesson made me nervous. Once we were told that we could pair up, that is when I felt relieved. Working with Mary was great, and I feel like she was more helpful than Mr. Franklin [the cooperating teacher who teaches high school technology]. She had creative ideas and we worked well together.*

Natalia commented that her cooperating teacher was "*very helpful with all of her "tips" and examples for producing a lesson plan and presentation. She provided great suggestions and insight for different tools to use to keep your students involved when teaching a topic.*"

All of the participants reported that the professional development workshops conducted by STEM professionals were beneficial. Many of the participants, although high-performing STEM majors, shared minimal belief in their ability to teach the subject. Across all three cohorts this lack of confidence emerged in different ways:

- Some did not feel they had enough knowledge and expertise to teach someone else the subject. "*Science is a very broad area and attempting to teach it is fairly difficult. It requires an in-depth understanding of the subject and the ability to teach it more than one way*" (Christina). There were impressed when observing in the classroom: "*She knew the topic that was being taught like the back of her hand, and she was able to efficiently relay that information to her students*" (Lily).
- "*How to properly relay information to students,*" "*Being able to communicate with children effectively, teaching them advance lessons without over teaching*" (Survey participant).
- Although comfortable and knowledgeable about the subject as a student, they were unsure how to explain concepts to someone else. How to teach using an inquiry approach. Although they were familiar with the framework, they wanted more guidance to complete several STEM-related activities during the PD, and when preparing for their presentation, they were not sure how to start. Survey responses included: "*I'm not a great teacher, I never know how to explain things is a way for others to grasp.*" "*In my experience, . . . I don't think I have the talent for teaching, because in most cases, it's hard for me to explain my point or to bring my point across in a way that children will understand.*"

Professional development

TIPS conducted a series of pedagogical workshops covering STEM content and included teaching participants to code, understanding the history of New York from multidisciplinary perspectives, geologically, geographically, biologically, and socioeconomically and exploring honeybees as biological and social organisms. In addition to STEM content knowledge, TIPS participants learned inquiry-based methodologies to actively engage students in the classroom.

The PD activities played a significant role in enabling participants to feel confident in their ability to teach, shifting from being high-achieving STEM majors to seeing themselves as teachers. A participant noted, "*the PD workshops informed me how to become an educator. . . ., the program has helped me with my confidence.*" Another said, "*I'm more aware how to use the skills from the PD sessions to teach . . . I learned how to structure lesson units . . . and I feel more comfortable moving along as a teacher.*"

Over time, we became intentional about adding STEM professional development (PD) sessions into our program, even amidst the global pandemic. Our PD sessions were usually facilitated by two STEM experts to maintain an ongoing mentoring relationship with the participants: a biochemist and a botany instructor from our local botanic garden. Each year the activities change and focus on a specific theme that helps STEM majors think about ways to engage K-12 students in hands-on inquiry-based learning. For example, the participants in Cohort 2 received care packages from the botanic garden, which allowed them to conduct hands-on experiments; they used lessons from the PD sessions to create a lesson that they will likely teach to students.

During the spring semester, we worked closely with our local Botanic Garden along with a biochemist who teaches in a Northeastern University in the Department of Biochemical Engineering to create a rich inquiry-based PD series. The Botanic Garden provided care packages, which contained soil, seeds, and informational text about gardening; this enhanced their experience and allowed for hands-on experiments during our synchronous and asynchronous sessions. Students engaged in botany lessons and learned how to bring nature into their homes and neighborhoods.

Our partner, a biochemist professor, developed innovative honeybee-themed PD workshops (see Appendix C). The PD workshops focused on gaining insight from honeybees to enhance teaching and learning. The five Honeybee PD Workshops focused on themes related to (1) Honeybee Flight, (2) Honeybee Stings, (3) Honeybee Honey, (4) Honeybee Communication, and (5) Honeybee Future Prospects and Projects. Participants found these sessions particularly engaging and creative, since this was the first time they were learning about the honeybee in a thematic structure. The focus was on empowering participants, as teacher candidates, to take an active role in designing inquiry-based learning modules and resources that will inspire transformative learning in their classroom. Our hope was that this process would be instrumental in enhancing their own personal learning and fueling/energizing higher levels of inspiration and passion for learning that would flow outward to their students and colleagues.

Participants noted that having these kinds of professional development were engaging and had numerous benefits. They got the chance to showcase what they learned from the PD series by presenting a lesson to our group during the last session. This process allowed them to plan and implement a lesson that they found useful since they did not have prior pedagogical experience. Concrete applications were beneficial to our STEM majors, whose confidence increased after teaching a STEM lesson.

One of the key findings from Cohort 2 was that participants needed such a program earlier in their academic careers. Participants felt as though having a program like TIPS should start during their sophomore year so that they have at least two years of programming. In addition, they wanted more pedagogical tools since they do not take education classes. To that end, we made some structural and long-term shifts to our program. Through a collaborative process with the Schools of Education, Science and Liberal Arts, we revised the minor in urban education to include a specialization in STEM. The minor allows any STEM major to take education classes and gain foundational and fundamental knowledge, skills, and practice to become a teacher.

Our program attracts students who, at minimum, have a curiosity about teaching. Though aware of the challenges of being a STEM major, respondents admitted to facing challenges as STEM majors but were not dissuaded from pursuing science and felt confident about choosing a STEM major despite the challenges. At the beginning of Year 3, three of four participants planned on pursuing careers in medicine and 25% in allied health. One respondent said "*I have a passion for helping people, especially those with sickle cell.*" Another respondent said, "*I love helping people and want to become a doctor, no matter how hard the courses are.*" Several participants detailed a greater purpose which was to be medically helpful to family.

Many of them reflect on how to transfer their passion for helping others from medicine to education. By the end of a full year, participants either wanted to become teachers or are open to the possibility of becoming a STEM educator. In alignment with previous research, key factors that contribute to this shift includes, but are not limited to, (1) exposure to the real teaching practice and classrooms, (2) STEM faculty to introduce them to the program (3) varied PD emphasizing pedagogy, and (4) mentorship by in-service teachers and faculty. To be most impactful, we focused on structuring the program to provide hands-on exposure to teachers in classrooms, which students often rank as a crucial element of our program.

In her reflection, one student states "*When I heard about TIPS, I immediately jumped on the opportunity . . . the program would allow me to discover a different career path and a different aspect of STEM. Before entering . . . I never*

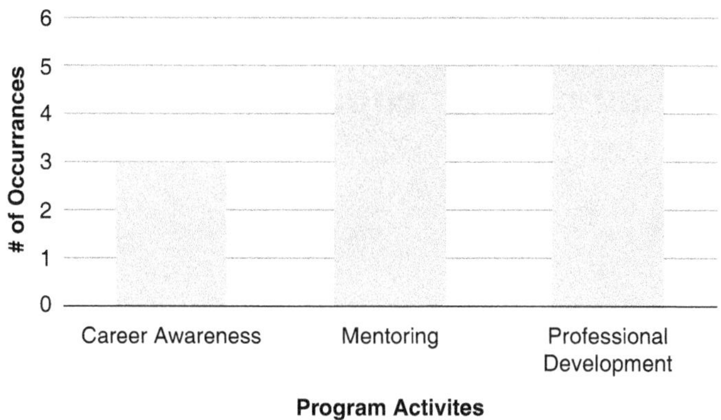

Figure 3.3 Program benefits reported by STEM majors

thought about becoming a teacher . . . However, since I did not know if I still wanted to go to medical school . . . this could be a potential alternative."

Our data suggests that TIPS participants reflect a group of high-achieving students who were confident in their ability to pursue STEM careers but less so of teaching as a career. Many of the participants in the TIPS program lacked knowledge of teaching or had not considered a teaching career in part due to lack of exposure and information. Over the year-long experience they developed more confidence in their ability to teach or share their content knowledge by explaining concepts to others. Their end-of-year written reflections show that career awareness, mentoring from peers, faculty, and cooperating teachers was perceived as a benefit of participating in the program. Mentoring and professional development occurred most consistently in the written reports (Figure 3.3).

Reference

Wallace, D. L., & Gagen, L. M. (2020). African American males' decisions to teach: Barriers, motivations, and supports necessary for completing a teacher preparation program. *Education and Urban Society*, *52*(3), 415–432. https://doi.org/10.1177/0013124519846294

4 Impact, lessons learned, and recommendations

Impact of COVID-19

The COVID-19 pandemic has changed the expectations and perceptions of students intending to enter medical professions and nursing. Our students tend to be nontraditional, working moms who are carrying a lot of familial and financial responsibilities.

There is a shortage of healthcare workers and a decrease in the number of students entering the nursing program. We have noticed a shift in participants' engagement and the number of students who are interested in our program. As we grow, we would like to provide an additional educational and pedagogical foundation for our students and integrate the Urban Education Minor into the program, starting with the students' sophomore year. In addition to the healthcare shortage, the COVID-19 pandemic has created a severe teacher shortage. To that end, there are schools that are willing to hire unlicensed teachers in a residency model that allows students to earn their degrees while working. For the upcoming year, Project TIPS will partner with local schools to hire our graduates to add a direct benefit to both the participants and the local urban schools that are experiencing a shortage of STEMS teachers.

In addition, participants were not able to do in-person observations because of the pandemic, and they expressed grave disappointment at this loss opportunity. Given the uncertainty brought by our new reality, we discussed new opportunities to have our pre-service teachers work with in-service teachers remotely. In contrast to participants in Cohorts 2 and 3, Cohort 1 benefited from visiting the classrooms and seeing classroom instruction. Due to the COVID-19 global pandemic, participants in Cohort 2 worked virtually with STEM teachers to design lesson plans, and Cohort 3 received PD from STEM teachers to support their lesson planning.

After the first year of TIPS, we continued to plan real classroom experiences for subsequent cohorts. These experiences were relevant to our students, who did not have prior pedagogical experiences. Unexpectedly, March 2, 2020, was the last day we worked with our participants in person, as all schools in the

DOI: 10.4324/9781003414452-5

district switched to remote learning on March 15, 2020, because of the global coronavirus pandemic.

Before the shift to online learning, participants worked directly with Ms. Wave in her eighth-grade living environment class. Like Cohort 1, participants in Cohort 2 found working in a real urban classroom valuable; they gained crucial pedagogical skills by observing how the teacher delivered lessons and offered summative and formative assessments to her students. While participants were familiar with using quantitative data, Ms. Wave provided feedback to her eighth-grade students and discussed how to use quantitative data. TIPS participants were able to take a familiar concept and understand how to use this work when working with students.

Cohort 3 was the first to have a full year virtually, which was a direct result of the global pandemic. Our university's student teaching policy for the 2020–2021 academic year was for all experiences to take place virtually. Therefore, our professional development series and field-based opportunities were held synchronously and asynchronously. One of the key lessons learned from Cohort 2 was the need for more concrete pedagogical tools. Participants wanted materials that they would be able to implement into any lesson or unit they planned and implemented. To this end, we read peer-reviewed text related to curriculum development, varied pedagogical approaches, which work best for diverse student populations, and nuanced teaching styles. In addition, participants viewed videos of highly effective STEM teachers and analyzed the lessons.

Lessons learned

Methods used for self-study

For an in-depth examination of the TIPS program—one that documents participants' voices and experiences in a naturally occurring context—we used a broad range of qualitative methods such as self-study, narrative, ethnographic, case study, action research, and phenomenology to learn about what's working and what should be improved in the program. Merging these techniques allowed us to include multiple perspectives of program participants at different touchpoints in the program. Furthermore, it provided us with a mechanism through which we can replicate several program aspects to make comparisons across cohorts. As participant observers in the program, we also relied upon ethnographic, self-study, and narrative methods to reflect on our own experiences, beliefs, decision-making, and insights gained as the program evolved.

By focusing on micro-level practices to develop detailed descriptions, practices, and activities (Geertz, 1973), we used ongoing data collection to guide our design and decision-making to support our participants' needs. This kind of narrative inquiry and action research was responsive, timely, and organic.

It allowed us to draw from our experience to better understand contextual factors and local tensions and then develop theories about effective strategies (Tight, 2017) that can be implemented to address those issues. By studying our work within a local context, we could document individual experiences and perspectives as well as draw conclusions about the connections between different variables (Tight, 2017) as we considered the relationship between program activities and participants' outcomes or choices.

We also used case studies developed based on participant experiences with the program (Mills et al., 2010) to create profiles that can help inform our recruitment efforts. We will share later how developing a better understanding of the lived experiences of the students in the TIPS program guided when and how we used targeted recruitment strategies that utilized faculty and peer-to-peer methods to inform STEM majors about teaching as a career choice. Additionally, the case study of our program helped to inform the project through diverse evidence—specifically ethnographic and narrative data—as it occurred naturally (Tight, 2017). The descriptive approach of the case study method helped us closely examine the practices of a small sample of participants through in-depth study within a specific context (Becker et al., 2012; Rashid et al., 2019) to provide insight into their characteristics and experiences.

Because we planned to replicate the program and continue it in subsequent years, the co-directors and co-authors extensively documented the design and implementation processes through field notes, surveys, semi-structured interviews, and group meeting notes. We also used self-study methods to understand our own decision-making as institutional agents.

We continually reviewed program data to make improvements and shared our individual perspectives on program activities to draw conclusions about lessons learned. Self-study research is focused on looking inward to understand one's practice through systematic and objective examination of practice, particularly in teacher education, to improve individual practice as well as practices in the wider field (Hauge, 2021; White & Jarvis, n.d.). It allows us to remove the silos that tend to make teaching isolating to slow down practice and take a close look at the interactions, factors, and decision-making through an objective lens (Hannigan et al., 2016). Aligning self-study with case study allowed us to go deeper into our examination of practice based on the different disciplinary perspective we brought to the study.

Our work was guided by several questions, which emerged at different stages of program design and implementation.

- What factors impact recruitment of STEM majors?
- What can help address barriers to increase interest in teaching?
- What lessons does a collaborative STEM education program teach us about recruiting Black STEM majors?
- What support do Black STEM majors need to become classroom teachers?

- What kinds of experiences shape STEM majors' decision-making when determining whether to pursue teaching?
- What factors do high-performing STEM majors of color consider when deciding to pursue teaching?
- Which aspects of the TIPS program do STEM majors of color perceive as beneficial to their decision to transition into teaching?
- What types of field-based experiences do STEM majors with no prior classroom experience need to prepare for a career in education?
- What lessons can we learn about recruiting and retaining Black STEM majors to become teachers from partnering them with STEM field experts?

These questions helped us explore and test ideas throughout the project to see which approaches were most effective. We also used these questions to learn more about our participants. As we grew to better understand program participants, we were able to make adjustments and be more responsive to their needs and concerns as students and to their professional goals. These questions also helped us examine the program in depth and from multiple perspectives. We were able to identify gaps in the activities being offered and limitations in the data collection methods and tools being used to measure and learn more about what's working and what's not.

Over the year-long TIPS program, we collect numerous data sources. Throughout the academic year, we hold group meetings—which can range from 45 minutes to an hour and a half. We conduct focus groups twice a year, conduct semi-structured interviews at the end of the year, and collect program surveys at the beginning and end of the year. The survey is an 11-item interest questionnaire to learn about the participants' career interests, perceptions of teaching, likelihood to pursue teaching as a profession, and perceived needs should they transfer from STEM major to STEM educator. We use focus group meetings each semester to learn participants' perspectives on program activities and to identify areas that can be added to and/or enhanced in the program. During our meetings, we take copious field notes to recall pertinent information and comments about potential adjustments needed for the program. We collect and review our meeting minutes as well as written reflections from the participants. At the end of the year, we ask participants to submit a written reflection on their overall experience. They are asked to think about the program activities and how the experiences impacted them and their decision-making.

Over the years, we have made numerous changes to our data collection methods and tools based on limitations and gaps we recognized early in the program. Initially, we did not collect demographic information about participants at the onset. We also did not survey participants to learn about their prior knowledge or teaching experiences.

In Year 1, much of the data collection procedures were summative: interviewing participants, collecting end-of-year reflections and program surveys, and facilitating focus groups after specific learning experiences to gain insight. We held program meetings bi-monthly; they were informal meetings to get caught up and plan activities for the participants, to learn about them, their interests and needs, and to share information and support. The program surveys captured information about participants' interest in teaching and their needs moving forward after completing the TIPS program.

In Year 2, we administered surveys as a formative assessment to collect information upfront from program participants, which we then used to plan monthly meetings. We collected the formative assessment, a Google Forms survey, prior to any planned activities to learn more about the career goals of STEM majors and the factors influencing their decision to pursue or not pursue a career in education.

During Year 3, the mandatory shift to online learning during the spring 2020 semester prompted us to conduct asynchronous and synchronous PD sessions. We incorporated pre- and post-professional development surveys based on the content areas and topics discussed in the workshops to gauge how the professional learning was impacting the participants' pedagogical content knowledge.

Initially, TIPS participants received a stipend of $500 for their participation. During Year 3, to increase student participation, we increased the stipend for participation to $500 upfront and $1,500 upon program completion. Our decision was based on available funds and the acknowledgment that most of our participants are working students, oftentimes with full-time jobs. We wanted to add value to the experience for students by providing increased financial support.

According to Bowen (2009), using an audit trail is advantageous because it combines the research process with program activities by documenting methods through detailed record-keeping. We take copious notes at every meeting and program-related activity, namely focus groups, field visits to classrooms, interviews, workshops, questions from participants, mentoring meetings, professional development experiences, and academic advising sessions. The notes include comments, reflections, and memos to create an audit trail (Bowen, 2009) of the project and to facilitate decision-making that guided the program.

The open-ended nature of the TIPS program—designed to meet the needs of each cohort—required a grounded theory approach to address rigor and validity (Bowen, 2009). Following this approach, we used iterative and interpretive protocols to examine the data and generate hypotheses throughout the process so that we could make program corrections. Program meetings increased in frequency and eventually became weekly meetings during the first part of Year 3 and then bi-weekly, to help enhance program activities and provide time for in-depth self-study. We shared our individual field notes and perspectives during planning meetings and used them to make programmatic decisions. We spent a lot of time reviewing and discussing our data, program goals, and the broader implications of the program for teacher preparation, STEM education, and preparing teachers to work with diverse learners in urban, low-income schools. When reviewing our data, we initially used an iterative process. We looked at each source individually to identify emerging themes. Through a constant comparative, recursive process, we shifted through the data to corroborate the examples emerging from each source. We discussed the patterns we noticed and grouped similar examples into categories. Then we contextualized these patterns within the broader program to triangulate and confirm our findings.

Thematic analysis of qualitative data revealed several patterns in the data: motivation to join the program, uncertain career path, plans/planning, self-perception as a teacher, efficacy/confidence/knowledge, benefits of the program. These were initially grouped into three categories: professional development, mentoring, and career awareness. Over time, recursive analysis of the qualitative data identified the frequency with which each of these categories emerged in the written reports. At the end of each semester, these findings were triangulated with the survey results to develop conclusions about the efficacy of the program. We returned to the findings and regrouped the data into two categories: concerns about teaching as a profession and exploration, mentoring, and professional development. This allowed us to discuss the results to highlight the ways in which the TIPS program addressed the participants' concerns through a cause-and-effect lens.

These qualitative methods created an audit trail because we documented our practices chronologically (Bowen, 2009), which allowed us to understand the project over time (Tight, 2017); therefore, we created replicable, systematic processes that we could use to enhance and expand the program each year. In addition, through our collaborative inquiry we engaged in reflective self-study as a learning community, looking specifically at our individual contributions to the overall program to improve our learning and practice (Sampras, 2011).

Our self-study and program evaluation revealed several factors and lessons learned that appear to influence STEM majors when deciding whether to pursue teaching. The findings from our inquiry focus on two areas:

1. Recruitment: Recruiting STEM majors to become teachers is aided by mentoring, in-person classroom experiences, and deliberately targeting second-year students (sophomores). STEM majors consider several factors when deciding to transition into teaching. The most significant factors appear to be familial obligations and pressure; their prior knowledge and experiences with education, specifically schooling in urban contexts; and financial benefits of the teaching profession. STEM majors believe that mentoring and opportunities to work directly with students in real classrooms were beneficial. More specifically, mentoring by faculty and cooperating teachers who shared similar career pathways, backgrounds, and experiences helped provide the social capital to support STEM majors as they transitioned professionally from STEM major to STEM education.

2. Preparation: Real-world opportunities to work with diverse learners influenced STEM majors' perceptions of themselves as potential teachers and helped them think more about teaching as a profession. Participants found working with teachers in real urban classrooms particularly beneficial, as these experiences countered many myths and preconceived ideologies the participants previously held.

Below, we discuss these findings, share specific details, and have in-depth discussions about program activities and participant outcomes. We also provide recommendations for other PBIs as well as institutions serving a broad array of STEM majors who can potentially enter the teaching workforce.

Program design

We drew upon the research to design a program that would support STEM majors of color. Consequently, our program design was not a linear process but an organic one that took an action research approach. We were continuously guided by feedback and data from our participants to inform our decision-making. As a team, we asked a myriad of questions about the students we served and what institutional structures appeared to pose barriers to their success. With each new cohort, the more we met with the students in the TIPS program and tweaked and adapted our activities, the more we were able to confirm that Predominantly Black Institutions (PBIs) and Historically Black Colleges and Universities (HBCUs) are strong potential sources of candidates when recruiting teachers of color (Wallace & Gagen, 2020), but institutional barriers prevented some of the possibilities. We looked closely at our work to determine ways in which we can create processes to strategically recruit prospective STEM teachers. We came to realize that understanding the context and how we utilized local resources were important parts of the process. Below, we share insights we gained through this project and the impact on participants.

Identification and selection of participants

Initially, the required GPA to participate in the program was 3.5; this was later reduced to 3.0 for their overall GPA. Although most students average 3.2 GPA and above, this change was intended to recruit students who were more likely to switch to teaching and widen the pool of applicants to those who have high GPAs in STEM-related subjects but may have a lower overall GPA. Prospective students had completed an application.

Rather than using GPA as a barrier, we shifted focus to emphasize social support, pedagogy, where they were in their career planning, and how the program can help increase their interest in teaching. We examined how the TIPS program helps STEM majors as they considered teaching and focused on certain factors that appear to be influencing participants' decision-making, such as their understanding of teaching as a profession, their perceptions of themselves and their ability to teach, as well as cultural and institutional factors. We found that three factors helped to shape and explain participants' decision-making as they move from STEM major to STEM educator: their identity shift from STEM major to STEM educator; mentoring and information so they can make an informed decision; and providing concrete learning experiences to help build their efficacy and background knowledge about pedagogy. Participating in the program allowed STEM majors to explore teaching as a plausible career and address their concerns. They built their background knowledge about the field and reshaped their perceptions about teaching in urban schools.

Given the urgent need to recruit STEM teachers prepared to work in high-need urban schools, teacher educators must consider the resources and opportunities that facilitate the transition of STEM majors into teaching. STEM majors of color must reconcile several cultural factors that influence their decision-making, and social support such as mentoring can help them reconcile the tensions (Gist, 2017). The local context plays an important role in this process. For example, when looking at the local population of Black and Latinx migrants, New York and Florida were the two states with the most Caribbean immigrants between 2013 and 2017 (Zong & Batalova, 2019). Between these years, New York was also the destination for the second-largest group of immigrants (9%) from Sub-Saharan Africa (Echeverria-Estrada & Batalova, 2019).

Mentoring

In the context of PBIs, mentoring by college faculty of color to STEM majors of color who are transitioning to teaching is a significant factor in recruitment (Lawrence et al., 2019). We have learned that we must prioritize mentoring, specifically for STEM majors of color who bring cultural identities that might pose an obstacle to transitioning to STEM education (Lawrence et al., 2019). When provided with tangible support, consistent encouragement, and resources, TIPS participants' self-efficacy improved. Students noted that they were grateful for the opportunity and felt the program provided exposure they would not have had otherwise. In explaining their reasons to join the program, students emphasized the desire to shift careers.

Many TIPS participants needed to learn more about the teaching profession when they joined the program, which had implications for their decision-making. First, many participants did not consider teaching as an option because it was not on their radar, primarily because of lack of exposure or access. A survey respondent said:

> *I never had any exposure. No one from my entourage is in that field [education] so I don't know much about it. Also, I never really considered a career in teaching because my goal was set on something else. I always knew that I wanted to work either in the medical field or work in something that involved helping others. A teaching position never came to mind; however, I am open to it.*

Others had preconceived notions about teaching and students in urban schools based on their own experiences. When asked about concerns around becoming a teacher, one person noted: "*I am also scared of children [*sic*] attitude towards teachers since I have seen some very rude children.*"

Several TIPS participants had never considered teaching as an option. As institutional agents, college faculty and cooperating teachers in the partnering middle school where TIPS participants completed fieldwork provided insight and support around navigating educational institutions. For example, monthly on-campus TIPS workshops focused on advisement, pathways for transitioning to teaching, and facilitating peer-to-peer interactions. Over the past three years, faculty shared

- reasons why they became teachers;
- processes and pathways they used to become teachers;
- pitfalls and challenges encountered as teachers;
- benefits of being a teacher in urban schools; and
- financial resources for graduate study.

When we take a closer look at program practices, we see that social capital played a strong role when STEM majors of color decided (or not) to pursue teaching. Individuals who have walked the path you are about to embark upon provide a different perspective and resource. This kind of mentoring allowed the participants to envision themselves in the profession by hearing about the road college faculty and cooperating teachers had traveled.

Cooperating teachers also provided valuable support for prospective teachers (Matsko et al., 2020). Participants benefited from the additional layer of working with cooperating teachers who share similar backgrounds and academic experience. The cooperating teachers at the partnering middle school also provided mentoring; they were both nontraditional teachers who had transitioned to teaching from STEM.

All participants reported that the classroom observations were beneficial. Evelyn said in her reflection, "*Beginning this venture, outside of my comfort zone, was needed because it gave me perspective on what else can be a possibility, as a future career.*"

They also reported that the opportunity to speak with the classroom teacher was important. Daisy connected with her cooperating teachers as mentors. She wrote:

> *I enjoyed listening to the stories of Ms. Wave and another science teacher on how they went through the process of becoming teachers and what influenced them to choose the career path they did. Like myself, teaching was not what either of them thought that they would be doing. Both teachers were on the path to medical school when they decided to give teaching a try. By doing this, they both realized that they had a passion for this newfound craft and this shined through in both of their classrooms. I also enjoyed learning about first-hand experiences with programs such as New York Teaching Fellows and Teach for America, which both women participated in order to become educators.*

Daisy saw parallels between her pathway into the profession and that of the science teacher she observed. Her comments indicate that she benefited from learning that the mentor teachers had entered teaching through alternate routes beyond traditional teacher preparation programs.

During monthly workshops, as noted above, faculty of color who had similar backgrounds as the TIPS participants (Lawrence et al., 2019) shared insight about their own pathways to teaching; they often reflected on familial conversations, learning to navigate higher education, and making the career moves to become educators. We'd hope to recruit faculty across STEM disciplines to expand the resources available to participants who minored in subjects such as math and environmental science. One challenge we encountered when attempting to recruit more faculty was their availability and time commitment. Funding was not available for faculty stipends, which posed a challenge when recruiting mentors.

We also found that the need for and value of peer-to-peer mentoring and support were evident during our interactions with the participants. This insight facilitated a shift in our recruitment strategies. During Year 3, drawing on our observations of peer-to-peer mentoring in the program, we asked participants to bring a friend to the information session, and we recognized the importance of faculty. During the session, the STEM professor highly recommended the program and reinforced the value of the TIPS project by reminding students that they would receive extra credit for participating. This also revealed the significance of faculty commitment and buy-in. The true asset is the partnership between the STEM and education faculty coordinating the program.

The fieldwork ranged from classroom observations, assisting in the classroom by working in small groups, working in pairs to develop an inquiry-based lesson plan they would teach to students, and preparing and presenting on a STEM topic. Professional development workshops were led by STEM industry professionals on a wide range of diverse STEM topics, including coding, biological, and social importance of honeybees and the geological, historical, and environmental evolution of New York City.

The program benefited participants by increasing their understanding of teaching as a profession. For some, this required dispelling myths about teaching and students in urban schools. For others, this experience informed them of a career pathway and facilitated real-world experiences to help provide a frame of reference for STEM teaching in urban schools. Observing in the classroom was most impactful on their professional development, confidence, and identity shift to teaching. The students benefit from a cooperating teacher with similar content knowledge to help them understand how to teach the content. Making a leap to an integrated planning experience requires another skill level. They benefited from peer-to-peer support within the network. Mentoring benefited the participants because they were able to learn from more experienced teachers as well as peers who had some prior experience tutoring or working with children.

Despite programmatic shifts year after year, consistent mentoring and exposure to teaching through professional experiences helped to increase awareness and identify some core elements for replication. First, the program model should be used with STEM majors earlier, preferably first-year students or sophomores. This will give students more time to process a shift in career direction and be knowledgeable of their options. Second, college courses in education can provide STEM majors who already express an interest in teaching with the opportunity to obtain a minor in education while completing their STEM degree.

As the TIPS project grows and evolves, it continues to teach the authors numerous lessons. Cohort 1 made a request to have more professional development, a move that Cohort 2 deeply appreciated as PD became central to candidate's experience. As we began Year 3, we noted that candidates needed additional pedagogical support given their lack of exposure to the classroom. Since we were not able to physically be in classroom spaces, video analysis and scholarly materials become essential tools to introduce participants to the classroom. Providing pedagogical materials to candidates as we mentor them took on greater emphasis in Year 3, because students were again able to work with a science teacher as in previous years.

In the first year of TIPS, we recruited 12 students, of whom four continued to the end of the year. Participants are required to complete 20 hours of service learning each year; year one proved this to be a central benefit to students. TIPS partnered with a neighboring middle school located centrally in an urban community, where participants worked with a highly competent science teacher to learn more about planning, instruction, and student assessment. Completing service hour within a real urban classroom proved to be crucial to shifting participants' perception of teaching. As previously stated, students held misconceptions about what to expect in urban classrooms, these false ideologies were dispelled once participants had the chance to work with a real science teacher. Participants noted that after being in Ms. Wave's class, they felt more confident about working in urban schools. They noted that students were engaged and excited to learn science. While some thought students were "rambunctious" being in Ms. Wave's class offered a new perspective—students in urban settings enjoy science when taught by a caring, warm demander teacher. Ms. Wave's class was one that was vibrant, fun, and a safe space to question and explore.

All participants noted that having a real classroom experience made them feel more adequately prepared to become teachers and continue the process to gain teacher licensure and formal education. Some areas of improvement that participants noted at the end of the first year were (1) more professional development from field experts and (2) creating concrete roadmaps that outline the various paths candidates can take to become fully certified teachers. Participants wanted more practice with interviewing for teaching positions and preparing demo lessons, as these are not lessons they receive as STEM majors.

We noticed two groups of participants: those who had prior experience in a "teaching" role and those who did not. Those with experience include babysitting, tutoring, mentoring, working in an afterschool program, or paraprofessional. Despite prior experience, the participants identified knowledge gaps and expressed apprehension working with students in a formal setting. One survey respondent said, "*I tutor 1 ore* [sic] *2 students in math and I always thought I never explain it the best way I could.*" Another survey respondent said, "*I'm not a great teacher, I never know how to explain things is* [sic] *a way for others to grasp.*" A third respondent from Year 3 shared, "*In my experience in teaching, generally I don't think I have the talent for teaching because in most cases it's hard for me to explain my point or to bring my point across in a way that children will understand.*" Daisy, a member of the first cohort who had previous experience working with children, was "*nervous and . . . not sure of what to expect*" from the field experience. She said:

> *I was familiar with schools and policies; however, I was not too familiar to how teachers would operate in the classroom during the day, and how interactions with middle-schoolers would differ from children who are in elementary school.*

We learned from STEM majors that they hesitate to pursue teaching primarily because of the way they or their families perceived teaching. Given our population of ethnically diverse Caribbean students, we found that family members pushed participants to join the healthcare industry. This was true for both Year 1 and Year 2 participants. One participant, Cee, had a strong interest in math and shared this with her family. While she had dreams of becoming a math teacher, her family encouraged her to find work in the medical field instead. This dampened her overall career confidence as she felt as though she was unable to pursue her true passion. Field notes from one of the co-directors show that during an interaction with Cee, when asked about her plans for employment and teaching, she dropped her head and expressed uncertainty about her career choice.

Cee, who was a single mother, instead focused on the work she was doing in lab at the Science Department, as she needed money to maintain her livelihood. Might her confidence have increased if she had the familial support to pursue employment as a math teacher, as opposed to seeing it as a badge of shame?

As we delved further into investigating the factors that inhibit STEM majors from pursuing careers in teaching, we found that in many ways teachers from their past indeed played a role in their dissuasion. Participants overwhelmingly expressed that prior to attending college they had negative experiences with STEM classes. Remy, a female participant who attended urban schools, noted that her STEM teachers never "looked like her and could not relate to her." She felt misunderstood in school, even though she demonstrated interest and excelled in STEM.

Frankie noted that her teacher did not encourage her to become a teacher; in fact, she felt as though the teachers complained about students' work, and teachers made it seem as though "they were only showing up for a check," and often discouraged students from entering a career in teaching. Another participant noted that some teachers read from the science textbooks and did not explain the content; therefore, the participant felt disconnected and disengaged.

For the first two years, this process yielded about the same number of attendees at the information sessions. Year 1, about 12 biology students expressed interest by attending the information session, and six submitted applications to participate in the program. In Year 2, eight attended the information session and five submitted applications. In the current year, Year 3, we focused our efforts on increasing interest and applications. We expanded recruitment efforts, which led to higher numbers at the fall virtual information session where 57 students attended an information session for the program. These higher numbers were the direct result of faculty recruiting more students, which involved more courses in the biology department. The recruiting information shared in the sophomore biology class focused on a discussion about why becoming a teacher can be fulfilling, providing an opportunity to support the next generation, continuing to work with the content they love, and how transitioning from a content perspective to a teaching perspective can be beneficial.

Each fall the TIPS program launches with a new cohort of students (Table 4.1), and retention rates have improved over the past five years. In Year 1, six students started and four completed the year-long program (67% retention). Years 2 and 3 had 50% retention, which we believe was attributed to the COVID-19 pandemic and the shift to remote learning. During Years 4 and 5, all of the students who started the program completed it (100% retention).

All cohorts completed program surveys early in the program; these initial surveys showed that TIPS participants most commonly expressed concerns about apprehension about teaching, (lack of) teaching experiences, teacher salary, and family pressure, expectations, or demands. Familial obligations appear to a consistent factor impacting STEM majors' decisions to enter teaching. It is important to note that in the formative survey, 42.9% of the responses from

Table 4.1 Retention and outcomes of program completers

Year	*TIPS program completers*	*Post-TIPS participant outcomes*	*Percentage of TIPS participants pursuing teaching*
1	4	1 out of 4 participants became a high school science teacher 1 out of 4 pursued nursing 1 out of 4 enrolled in medical school 1 out of 4 pursued pharmacy	25%
2	5	1 out of 5 participants pursued science teaching 4 out of 5 pursued nursing and medicine	20%
3	5	3 out of 5 participants became science teachers in elementary and high schools 2 out of 5 pursued nursing and medicine	60%
4	7	2 out of 5 are current matriculates will graduate in June 2023. Based on current reports, 3 interested in teaching, 1 has pursued nursing, 1 has pursued dentistry, and 1 will pursue medicine	57%
5	7	This is the current year 1 student is pursuing the minor in education and will apply for teaching All students are still matriculated	N/A

Cohort 1, 28.6% from Cohort 2, and 28.9% from Cohort 3 indicate family pressure, expectations, or demands impacted their decision. One survey respondent said, "*My family wants me to be in the medical field so teaching was never an option. However, this is something that is very interesting*." The participant felt as though she had to abide by the family guidelines and that selecting any other option would be too hurtful to her family.

The participants' concerns and needs were consistent across cohorts. A snapshot of the responses from Years 1 to 3 reveals concerns about salary, teaching knowledge and skills, and the career path (Table 4.2). To overcome this barrier, we designed workshops to inform students about state certification requirements, teacher salaries, and the kinds of experiences and mentoring provided by teacher educators. To better understand these factors, we looked closely at a snapshot of the first three cohorts of the program.

Given the prevalence of these factors in the survey and our monthly meetings, for our analysis we looked at the career choices of the participants in Cohorts 1–3 (Table 4.3). We found that the interconnections of gender, race,

Table 4.2 Self-reported factors impacting decision to pursue teaching (Cohorts 1–3)

Factors	*Cohort 1 (N = 7)*	*Cohort 2 (N = 7)*	*Cohort 3 (N = 38)*
Teaching certification exams and requirements	28.6	57.1	34.2
(Lack of) teaching experiences	42.9	85.7	60.5
Teachers' salary	57.1	28.6	26.3
Family pressure/ expectations/ demands	42.9	28.6	28.9

Table 4.3 Pre- and post-career choices of TIP participants (Cohorts 1–3)

Cohort	*Career Aspiration before joining TIPS/ beginning of college*	*Career Aspiration after joining TIPS*
1	2 nursing 1 medicine 1 unsure	1 medicine 2 teaching 1 nursing
2	Hematologist-oncologist 4 nursing	Hematologist-oncologist 1 nursing 3 teaching
3	1 pediatrician 1 hematologist 3 nursing	3 teaching 1 pediatrician 1 hematologist

ethnicity, class, student aspirations, family habitus, and expectations appear to enable or hinder mindsets for STEM majors as they consider whether to become teachers. During Years 4 and 5, we made adjustments that considered these interconnections, so we can create more successful experiences, specifically around the mentoring topics discussed during meetings, that can help to shift perceptions, mindsets, and identity to improve STEM and STEM teaching recruitment.

Black STEM majors also face economic tension, particularly financial barriers that include having to work as they complete their degree to pay tuition (Rogers-Ard et al., 2012). For many Black STEM majors, this financial barrier poses a significant tension when they are making decisions about their career. One survey respondent said:

> *I am kind of undecided. Originally, I wanted to be a pediatrician, but I am also considering psychology. As of right now, I don't have a set position. Ultimately, my goal is to work in a field that will make me happy. I do not want work in a field because it is going to have a "good salary." Yes, a good*

> *a salary is very advantageous, but that does not necessarily mean that the job will make me happy. Think about it this way: if when you are at work and you feel like your job is stressing you, you don't like the environment that you are in, you don't get along with your colleagues, you feel like you keep working and never have time for yourself because whenever you get off of work you are tired, is this job really for you? Probably not. I want to work in a field that brings joy to my life. The biggest mistake that people make is that they tend to settle for what they have because they feel like they don't have any alternatives.*

To improve our programming, we developed profiles based on what we learned about participants, their requests around their needs, and what they believed would impact their decisions to teach: more information about teaching and working in classrooms; more content knowledge and how to teach content; and more hands-on teaching and skills (Figure 4.1).

Although much of the information about teaching, becoming a teacher, the salaries, and so on is publicly available, our presentations and workshops provided participants with insight on the career. Evelyn, a member of the first cohort, reflected on her experience in the TIPS program and said that participating in the program was "outside her comfort zone." Her comments suggest that when STEM majors transition to teaching, they must reposition themselves and move into uncharted territory. Evelyn's comments represent the professional and identity shifts many of her STEM peers feel as they transition into the role of STEM educator. As noted above, many of them did not see themselves as teachers.

Along with shifting mindsets, which begin with self-perceptions about becoming prospective teachers, cultural shifts also influence the career choices students make. The transition from STEM major to STEM educator can be

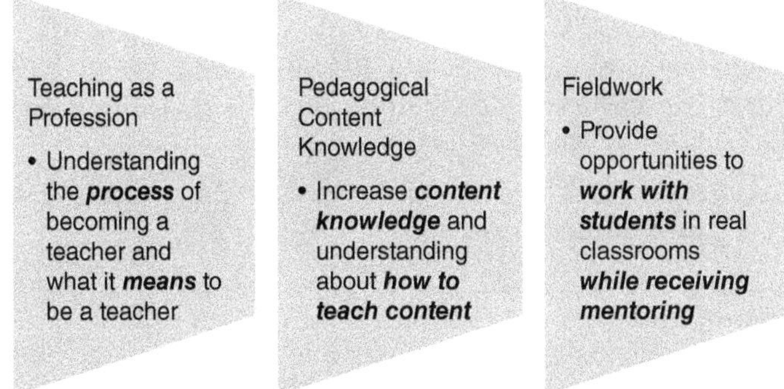

Figure 4.1 Components for TIPS program

quite jarring for some prospective teachers. Gist (2017) noted that "teachers of color make difficult choices concerning their commitments to the profession, communities of color, and students" (p. 927).

Most STEM majors in our program have been on a path toward the medical field. In formative surveys, respondents identified career goals such as marine conservation, medical doctor, nursing, software development, nursing or teaching, physician assistant, dental hygiene, or infectious disease specialist. Several people were not sure about their career goals.

Respondents across cohorts often shared that teaching was not part of their plan. One respondent said, "*Although I want to be a doctor, my mom (along with other family members) pushed me into the field saying how great I'd be as a doctor. Teaching has never been an option for me till now.*" Another said, "*I really want to pursue teaching. However, when I started college that was not the plan. So, I need to plan how to transition from what I'm currently doing to teachers' education.*"

For many STEM majors, these career goals are a big part of their identity. For some, their goals are linked to their family goals. Additionally, respondents expressed frequent concerns about learning how to teach, reflected in questions such as "*how do I learn the steps and strategies or what the right techniques to use [*sic*]?*" Some who expressed interest in the TIPS program said they did not consider teaching as a career option and others expressed concerns about how to teach their subject area, which also included trepidation about what to expect during field experiences.

Issues such as the familial role in becoming a teacher continue to be the key deterrent for our participant group, students who are largely immigrants and all of color. To many of the participants' families, teaching is not viewed as favorably as becoming a nurse or doctor. During Year 2, we learned that teachers and educators also cause participants to view the teaching in unfavorable ways. Disparaging comments about salary, workload, student behavior, and so on painted teaching in a negative light that made participants look the other way.

As we consider strategies to increase the diversity of STEM teachers, we must work with classroom teachers to examine their teacher identity and perception. We must have honest dialogue about the ways in which practitioners discourage students from joining the field.

Selecting mentor teachers

Purposeful and intentional recruitment of teacher mentors is just as important as targeted recruitment of students. The teachers can have a positive or negative impact on the experiences of the people considering teaching. Mentors that have a lot in common with the TIPS participants received accolades while others were not as effective. During Cohort 1, a TIPS participant reflected on the experience with her mentor teacher. She wrote in her reflection:

> *I enjoyed observing the class and student interactions, one thing that also I enjoyed was listening to the stories of [Ms. Wave] and another science teacher on how they went through the process of becoming teachers and what influenced them to choose the career path that they did. Like myself, teaching was not what neither of them thought that they would be doing. Both teachers were on the path to medical school when they decided to give teaching a try.*

In contrast, two reflections from Cohort 3 raised concerns about the lack of support received from mentors. One TIPS participant said:

> *My experience with [Mr. Calloway] could have been better. [Mr. Calloway] is a very nice person, but while we were working on our final project, I felt like he was not really invested in it. For him, since science was not his strong suit, he just left everything up to us and truthfully, that was kind of a disappointment. Most students who joined this program are actually considering working in this field, but since that is not my case, I had hoped for more support and guidance.*
>
> *There was a point where I was ready to quit and just get out of the program. I had no idea on what I wanted to present and not knowing how to present a lesson made me nervous. Once we were told that we could pair up, that is when I felt relieved. Working with Laurie was great, and I feel like she was more helpful than [Mr. Calloway].*

Another student said:

> *Not picking well prepared mentors that are willing and able to assist students in the program. I believe to thwart that issue is to build a system of accountability. Although, this would mean more stress on our TIPS program hosts (which I would not want), it is necessary. Having a system of accountability for the mentors should encourage them to be more eager to help.*

In an effort to increase the quality of mentor teachers' selected to work with TIPS participants, a protocol was developed.

Developing a sustainable program: considerations for recruiting Black/African American and Afro-Caribbean STEM majors

Over the years, we've refined the project to develop a better understanding of the factors impacting decisions of STEM majors of color who do and do not consider teaching. First, high-performing STEM majors may feel they do not have other career options outside of medicine that are available to them at their current institution. Second is increasing understanding about the teaching profession while allowing STEM majors to explore teaching as a viable option. In fact, we make the case for using career exploration as the basis for a recruitment program designed to increase STEM majors' interest in and awareness of teaching as a profession. Third, understanding the cultural influence of STEM majors of color and social capital are essential elements of teacher recruitment programs.

Through our research, we found the role that cultural and familial background play in recruiting STEM majors of color. In addition, given the lack of diversity and inadequate STEM subject area preparation that many of our participants received, we began to understand some of the factors that deter STEM majors of color from teaching. Two strategies appear to benefit STEM majors of color seeking to transition into teaching. First, support, guidance, and mentorship help retain and increase the percentage of STEM teachers of color through targeted recruitment of prospective teachers who share similar backgrounds as the diverse students in urban schools. Second, preparing teachers to work with diverse students requires learning more about racially, ethnically, and linguistically diverse students and working directly with them in STEM educational experiences.

Recommendation #1: broaden the definition of recruitment

Recruiting STEM majors into teaching is a long-term commitment. Previous strategies to recruit career changers through alternate route programs or residency programs fall short in their ability to foster STEM majors in undergraduate programs who have the potential to become teachers. Rather than just posting a flyer, TIPS uses a recruitment approach that helps increase knowledge and understanding of the teaching field and provides mentoring and field experience to build efficacy in working with kids.

STEM majors need time to understand how to shift their content knowledge into pedagogy while shifting their identities and self-perceptions about career goals. For example, PD workshops by STEM experts help STEM majors reconceptualize their understanding of STEM teaching and learning. It also helps increase and fill gaps in their own content knowledge. By the end of the program, most participants agreed or strongly agreed that the program increased their confidence in teaching (Figure 4.2).

Although some STEM majors have experience tutoring college peers or K-12 students, understanding the relationship between teaching, learning, planning, and building teacher–student relationships is important. TIPS participants who begin to perceive themselves as STEM educators when they develop confidence in their content knowledge and can explain concepts to students are the ones who consider teaching as a career option (Figure 4.3).

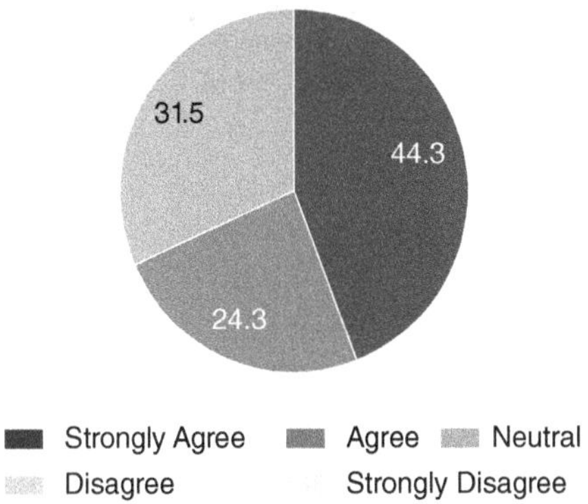

Figure 4.2 Confidence level of participants at the end of the program

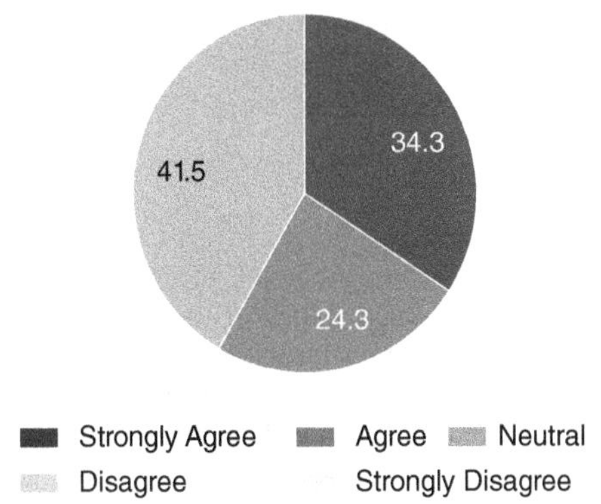

Figure 4.3 Participants' perspectives of mentoring and professional development activities

There does not appear to be any particular order in which TIPS participants acquire the attributes that indicate they are ready to teach (Figure 4.4). Some entered the program having experience working with kids but either needed to work on their confidence in being able to explain concepts or work with kids or needed to increase their pedagogical knowledge. Others had no background experience working with kids and needed opportunities to be in a supportive space to work with students.

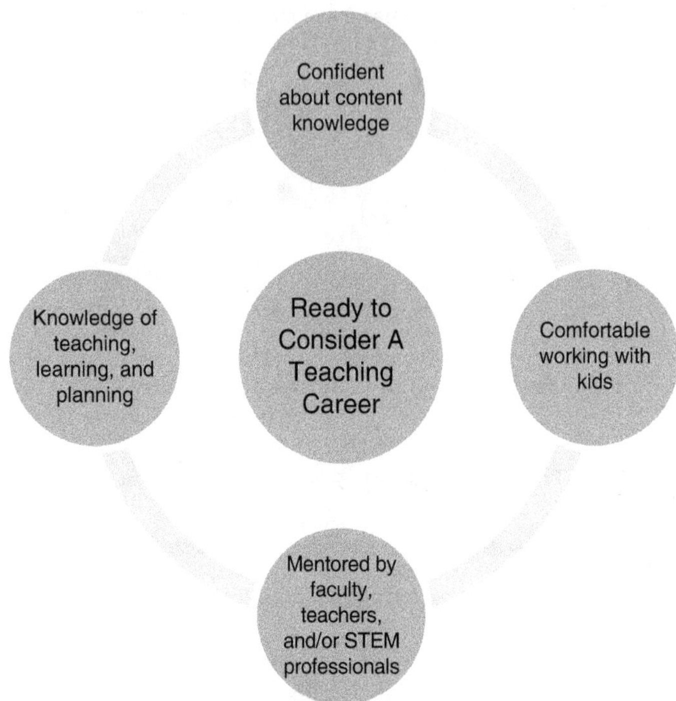

Figure 4.4 Attributes needed to teach

Traditional teacher education programs recruit and admit people who already know they want to teach. Much of their progression as they consider teaching is similar to traditional teacher education programs, but STEM majors perhaps never considered teaching. Therefore, introducing TIPS earlier in their undergraduate education can help create the identity shift sooner. During year 4, TIPS participants will have the option to pursue a minor in urban education, which begins their sophomore year. TIPS participants indicate that participating in the program increased their likelihood to pursue teaching (Figure 4.5).

Recommendation #2: target systemic barriers in educational opportunities

This project has identified a few areas for future program development and research. One challenge STEM majors of color need to overcome are the inequities students of color experience in academic spaces (Wallace & Gagen, 2020). Disparities continue to exist that perpetuate an achievement gap for

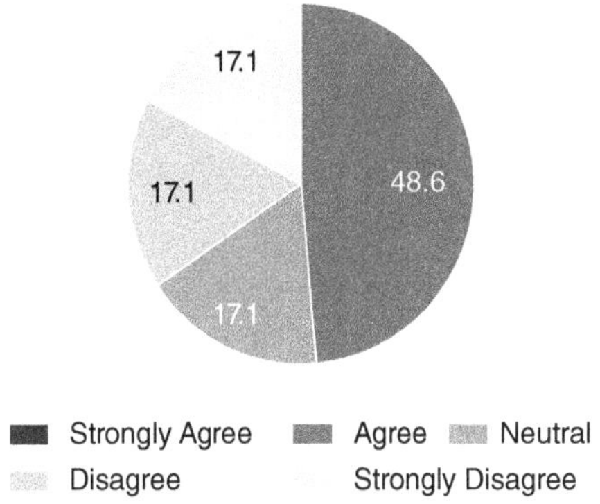

Figure 4.5 Likelihood of participants pursuing teaching after participating in the program

students of color, resulting in "the lack of males of color in the classroom" (Wallace & Gagen, 2020, p. 416) and a bleak pipeline for high-performing, diverse teachers.

To become a high-performing STEM major of color requires a certain amount of social capital, specifically mentoring, which is often overlooked in discussions about the influences on education. Additionally, career options may be limited for some students of color because of structural barriers as well as limited access to academic opportunities that would put them on a trajectory toward post-secondary STEM education. Consequently, whether pipelines and networks are abundant or limited, they influence decision-making. If we do not attend to these areas, we will perpetuate the current void that fuels a status quo grounded in stereotypes and dominant narratives that marginalize the education of black and brown students.

For many students of color, particularly STEM majors, social capital in the form of mentoring (Gist, 2017) can motivate them to persist through academic spaces that pose structural barriers (Samuelson & Litzler, 2016).

Social networks that include institutional agents can offer mentorship and access to institutional resources and opportunities (Stanton-Salazar, 1997). At institutions of higher education, particularly PBIs and HBCUs, STEM majors of color can gain access to mentoring by faculty of color. The problem, however, is the ongoing silo between higher education disciplines, which prevents students from gaining access to and information from institutional brokers who can assist and support their decision-making.

Although the previous frameworks attempt to critique social structures, by overlooking the cultural assets students of color bring with them to school, these frameworks perpetuate deficit views by minority groups.

A shift in perspective must include interpersonal networks that examine students of color in academic spaces because it is a "concentration on how social antagonisms and divisions existing in the wider society operate to problematize (if not undermine) minority children's access to opportunities and resources that are, by and large, taken-for-granted products of middle-class family, community, and school networks" (Stanton-Salazar, 1997, p. 3). Although some STEM majors are high-performing, they suffer from stereotype threats—stereotypes they hold about certain career pathways can hinder their level of self-efficacy and risk-taking. That is, they may be interested in teaching as a career, but few pursue this option out of fear of the unknown, of their perceptions of their content knowledge, and of their ability to teach that content.

How cooperating teachers are selected is also a critical area, alongside giving more attention to the kinds of field experiences and mentoring prospective STEM teachers receive. The cooperating teachers are practitioner teacher educators who help support prospective teachers and provide resources from a practitioner insider. Principals should give more purposeful planning and attention to how they identify cooperating teachers. In our follow-up research, we will look more closely at the cooperating teachers who work with TIPS participants, so these mentors can share their stories. Students of color need mentoring by people who share similar backgrounds and experiences (Lawrence et al., 2019).

One institutional shift in Year 3 was offering a minor in urban education, which provided STEM majors an alternate career option outside of medicine. With this option, STEM majors could take education courses during their senior year and the ensuing summer to complete a minor in education. Advisement records and student transcripts indicate that many STEM majors complete their content courses by the first semester of their senior year, and few of them have options for a full-time credit load in their final semester. Students typically take elective courses in an area of interest to fill their schedule for the last semester. Financial support from grant funds offered by private sector partners will cover 100% of the tuition for program participants who pursue the minor in urban education.

Recommendation #3: increase understanding of cultural influences

Culture may influence the decision-making of STEM majors of color who choose to transition into teaching. As noted in earlier chapters, local context matters. Most of the participants in the TIPS program are Black/African/African American females of Caribbean and African heritage. By examining how family and culture influence STEM student career choices, we reveal how cultural and family expectations reinforced positive STEM identities for TIPS participants. The positive reinforcement students received when they pursued nursing or medicine facilitated their high performance and persistence in

STEM. Meanwhile, cultural and family expectations cultivated negative perceptions of teaching that served as barriers to entry into the field.

Our findings help us understand the interconnections of three forms of capital—cultural, family, and STEM capital—that inform the mindset and choices of TIPS participants as STEM majors. We determine that getting students to choose STEM teaching requires three efforts: a shift of identity, reframing the perceptions of teaching, and reconciling the influence of family capital—transforming that influence from a barrier to a positively reinforcing factor for African, African American, or Afro-Caribbean women majoring in STEM.

To examine how family culture may impact students' aspirations and choices, we must consider the multiple factors that underscore this influence. These factors include students' ethnic backgrounds, whether students were born in the United States or foreign-born, whether their parents were educated or not, whether they feel confident doing STEM, and determining how students choose to navigate STEM careers. The key is to identify how the intersection of family and culture can positively influence and reinforce students' STEM identity but become barriers to teaching.

We seek to examine the intersections of culture, family habitus, and STEM identity at the college level for African American and Afro-Caribbean American students as they explore STEM education and STEM teaching career aspirations. This study arose from a collaboration designed to recruit biology majors to become STEM teachers in urban schools. Biology majors are recruited for a year-long experience that will expose them to teaching, to faculty and peer mentorships, to professional development, and to an apprenticeship in a middle school classroom with a licensed STEM teacher. TIPS participants have the option of completing a minor in urban education. In this chapter, we evaluate the following research questions regarding the influences of family and culture on STEM career choices:

1. What do female African American and Afro-Caribbean (underrepresented) biology majors report as factors influencing their decision to enter STEM and STEM teaching?
2. What role does familial habitus play in their decision to pursue STEM?
3. Does family and cultural capital, while reinforcing STEM identity and STEM-related careers in healthcare, create barriers to STEM teaching for TIPS participants?

Recommendation #4: focus on the intersections of culture, family habitus, and STEM identity

Multiple identities of STEM majors

Kim et al. (2018) sought to explore how aspects of the social environment influence STEM identity development. The qualitative research method was used to conduct this research since it is an approach that focuses on people and their experiences, behaviors, and opinions. The qualitative research method seeks to answer questions of "how" and "why," providing detailed insight and understanding. Their study found young women experience challenges to their participation and inclusion in STEM settings. In addition, they found that perceptions regarding who is part of an in-group or out-group of STEM fields can be changed through intervention in educational programs (Kim et al., 2018).

Ethnic minority students continue to exhibit diminishing interest in science and STEM participation; the consequence is underrepresentation in these areas. Many students struggle to embrace a STEM identity where they see themselves successfully engaging in STEM or choosing STEM careers. Most TIPS participants exhibited positive and well-formed STEM identities in pursuit of healthcare careers but were not confident in their ability to pursue teaching as a STEM teacher.

The complex identities of TIPS participants intersect not only to affect their motivations and how they navigate their educational journeys but also to create tensions around choices, especially around career choices. We can better understand how to have conversations around choices when we understand how these identities—as African American, Afro-Caribbean, or African women, as an immigrant in some cases, as someone from a lower socioeconomic class, or as the first generation to attend a college—work to reinforce or hinder students.

In a survey of sophomore biology majors in the classroom, we found only 6% of students considered a career as a STEM teacher. Given the small percentage of students who initially intended to become teachers, we were curious about whether students would change their mind after participating in the TIPS program. Hence, we evaluated factors that might impact the shifts or serve as barriers to TIPS students considering teaching STEM majors.

Influenced choices: nursing versus teaching

The TIPS participants were declared biology majors who were juniors or seniors. Before participating in the TIPS program, 57% of participants wanted to pursue nursing, 29% medicine, 14% allied health careers, and 8% teaching. Nationally, nursing students comprise more than half of all health professions students (American Association of Colleges of Nursing, 2020). Nursing is the largest healthcare profession in the United States, with more than 3.1 million registered nurses practicing nationwide (American Association of Colleges of Nursing, 2020). For many students, the decision to pursue nursing over medical school or other healthcare professions is seen as practical, secure, and attainable. All of the authors have family members in the nursing profession;

as early STEM majors, these family members were influenced by their family's history with and pursuit of the profession. It served to reinforce self-efficacy and academic abilities in pursuing STEM. To understand the choice of nursing as a primary career for students at Malcom College, we must examine the value Caribbean immigrant and African American families place on the profession.

Immigrants from the Caribbean have historically responded to the worldwide shortage of nurses in developed countries by emigrating to places like the United Kingdom, the United States, and Canada. In the Caribbean, nursing is considered a highly respected profession that is typically seen as a secure, well-paying job that allows families to transition from working to middle-class socioeconomic status. Therefore, it is not surprising that the experiences and expectations of immigrant parents living and raising their children in the United States may impact their children's career choices.

The value of nursing versus teaching

While nursing is perceived as a step below becoming a doctor for many students and their families, it is well respected and valued as a profession that helps people. Students feel empowered becoming a nurse. They feel less value in becoming a teacher. Although teaching is a respected profession, it is perceived as challenging, stressful, and not as lucrative as nursing, which we have detailed in an earlier study (Lawrence et al., 2019). This perception of teaching was influenced by students' own experiences in poorly managed urban classrooms and by general misinformation about school environments, workloads, and stereotypes about teachers receiving low salaries. When asked about their perceptions of teachers, students stated:

> "*Teachers don't make money.*"
> "*Students behave badly in the classroom.*"
> "*I don't know if I could control the kids; I might be too stern.*"
> "*Teachers are always stressed out.*"

We asked participants about their opinions of teaching as a career. One of the participants was Poppy, who was a 24-year-old immigrant from Jamaica and mother of two. Her children did not attend an American public school. Poppy said, "My teachers in Jamaica were very strict." From what she had seen on television, heard from family, or experienced at the public schools her children attended, Poppy thought that American public schools were terrible, and the kids were unruly.

When we asked other participants why teaching was not the first option for them, those who had attended public schools in New York City cited "ill-behaved students," "stressed-out teachers," and not being able to manage the students if they became a teacher. Their lived experiences played a role in their negative perceptions of the teaching profession.

These quotes seem to indicate a relationship between perceived value and resistance to teaching. Students perceive less value in teaching than nursing, potentially from their own experiences and likely through social and familial influences and stereotypes. Students do not see becoming a teacher as an attractive career option; they perceive the job as stressful and low-paying.

The student responses may be explained by expectancy-value theory. This theory posits that if students valued teaching, believed that they could successfully participate in teaching others, and perceived a low cost and high benefit to teaching, they would be more likely to choose to engage in teaching (Cooper et al., 2017). Because TIPS participants perceived a low initial value for teaching, it makes sense that they did not prioritize it as a career option. Other factors may also influence STEM teaching career choices.

Parental cultural capital: student versus family choices

A program survey of TIPS participants asked, "How much of your decision to pursue healthcare was influenced by your family?" One African American participant, Felicia, responded, "*Coming out of high school, nursing made money. I didn't really think about too many other career options*." She added she was "*not sure if to continue with nursing, especially during the COVID-19 pandemic*."

Another participant born in Jamaica, Cathy, said, "*My family wanted me to do nursing because they thought teachers . . . didn't make money*." She added that she liked math and decided to minor in it but never considered a teaching career as an option seriously. Brenda, a pre-med biology major, said, "*My family wanted me to be a nurse, but I decided to do pre-med and started the biology program*."

Althea was born in Jamaica but immigrated to the United States when she was a year old. She said, "*Mom wanted me to do nursing, but I thought I would be a doctor that worked with children . . . after completing the TIPS program . . . I had to tell Mom and Dad that I wouldn't be a nurse*" but would pursue STEM teaching. Similarly, a survey participant said, "*Although I want to be a doctor, my mom, along with other family members, pushed me into the field, saying how great I'd be as a doctor. Teaching has never been an option for me till now*." Overall, when program participants were asked: "What factors influence your decision to choose or not choose a career in teaching," 72% answered family expectations or demands (43%, Cohort 1, and 29%, Cohort 2).

These responses were very typical for many participants, highlighting how ethnic and cultural backgrounds along with family expectations inform students' complex identities. These identities are key factors in how students select a STEM major and in directing students' career path. We must consider these factors to understand how to shift from STEM student to STEM professional.

We sought to examine the experiences of STEM majors of color in a university-based teacher recruitment program. We learned that students who do not have immediate access to social or familial capital need support from more inter-departmental and inter-institutional partnerships; such partnerships can help students transition from STEM major to STEM educator. When we take an asset-based approach to program development and activities, we can help STEM majors of color better navigate these spaces. For example, supporting faculty of color and providing them with financial resources so they can work with students of color may transform the recruitment process for STEM education. More programs are also needed that support STEM majors of color with opportunities to tutor, mentor, and work with others throughout their undergraduate education. By partnering with other departments on campus for STEM majors to offer peer tutoring or partnering with local schools for service learning and afterschool mentoring opportunities, we can help to increase participants' content knowledge as well as hone their skills and efficacy.

Additionally, participants often lacked the institutional capital and networks to help them understand how to navigate the institution to facilitate their transition into teaching. In higher education spaces, therefore, college faculty play a key role in recruiting and mentoring STEM majors of color who seek to transition into teaching. The capital STEM majors gained when middle school teachers provided formal and informal mentoring helped the participants begin to reposition themselves and consider teaching as an option. Mentoring also helped them address personal tensions (Gist, 2017) from familial obligations as they considered teaching careers. It created space for shared stories about the various pathways into teaching and helped STEM majors affirm their decision to pursue teaching.

Many TIPS participants had pre-planned career paths based on familial goals. This familial capital emphasized and prioritized medicine as the career path for biology majors, specifically nursing for the predominantly female participants in our program. Although participants lacked knowledge of the field of teaching, what it means to be an educator, and the opportunities available through a career in education, family obligations appeared to be a significant factor in the participants' decision-making. Chapter 4 provides further insight on the role of culture by looking closely at the background and collective experiences of TIPS participants.

For STEM majors of color who have an interest in teaching, making the shift from a STEM career to a STEM educator required new perspectives on their identity. These new perspectives helped alleviate some fears about content knowledge and teaching in urban classrooms. Through mentoring and sharing stories about their paths into teaching, STEM and education faculty and middle school STEM teachers helped affirm STEM majors' decision to pursue teaching.

The participants' experiences in the TIPS program reflect several forms of capital. As established biology majors intending to pursue nursing, students exhibited STEM capital and STEM family habitus—the participants had positive STEM identities about their abilities in science and were deeply influenced by their families. This is not uncommon, as families influence how cultural norms and value systems are developed in their children. It is therefore not surprising that families also influence priorities at the tertiary level as it relates to STEM and career choices.

STEM capital reflects economic, cultural, and capital as factors that influence the shift that occurs for STEM majors initially directed into healthcare but who are engaged to consider a transition into STEM teaching. The TIPS network becomes a practical resource for students in this program because it provides the supports and networks students need to effectively use their education and cultural capital to gain economic capital. These networks include mentoring by faculty of color as well as peer networks consisting of individuals who share similar experiences and tensions (Gist, 2017).

TIPS participants use these unique capitals and diverse identities to make necessary shifts: from family member to individual decision-maker, student to professional, and working-class to middle-class members of society. They experience the tension of fulfilling their family's expectations and goals, oftentimes prioritizing those expectations and goals over their own. Furthermore, many STEM majors are unsure about future career options and goals beyond medicine. Many do not perceive themselves—high-performing STEM majors—as assets to students in high-need schools. As potential mentors and STEM teachers, they could influence the STEM experiences and shift the mindsets of the next generation. To shift their identity from STEM major to STEM educator, students must gain a new understanding and self-perception of how they can contribute beyond medicine.

When describing the benefits and limitations of networks, Bourdieu and Passeron's (1977) framework suggests that STEM majors might experience social conflicts when they shift from nursing to teaching. The professional development and peer and faculty mentoring experiences created by the TIPS program structure provided students with social relationships and networks that helped them leverage those systems to switch to teaching as a profession in a more informed manner. Still, many students continued along the healthcare track as initially intended, highlighting that their ability to choose is not made easier but more complicated by familial and cultural pressures.

After TIPS participants gained more knowledge about teaching and had experiences with teachers through professional development and in the classroom, some felt less resistance to teaching. These participants had a shift in value and opted to switch to teaching. However, many were still compelled to pursue healthcare careers after participating in the TIPS program. In total, 44% of TIPS participants switched from nursing/pre-medicine to pursuing teaching after completing the program. This is significantly less than those who would have pursued teaching on their own without participating in the program.

By considering several factors, the TIPS program serves to model best practices for recruiting more STEM teachers. Our findings illustrate the following three factors:

1. Family habitus and cultural and social capital are important resources in the process of STEM teacher recruitment. Family habitus is not only important in STEM at primary and secondary levels but at tertiary levels as well. Providing information to parents may be as crucial as the real-world experiences students gain; this information remains essential to dispelling negative stereotypes and misinformation about the teaching profession. We learned that African, Afro-American, and Afro-Caribbean families' influence impacts the professional career choices of the family member pursuing STEM to reflect their values.
2. We need to change the negative perceptions of teaching that students and their parents hold. Though TIPS participants embraced being good at science and participating in STEM, they must still overcome the stereotypes against teachers. Students must overcome the negative cultural perception of being a teacher compared to the positive perceptions of having a career in medicine.
3. Familial capital helped students overcome one barrier documented in the literature on STEM majors of color: they were able to persist and become high-performing STEM majors. At the same time, familial pressure, evidenced through mindsets and biases about teaching, shaped their perceptions and beliefs against becoming part of STEM education. To help students shift their mindset so they can see themselves as teachers, we must create a positive perception of teaching as a profession and reinforce the idea that STEM efficacy is a strength when students pursue this profession. For that shift to be effective, students must also be able to influence their family's perceptions and attitudes.

References

American Association of Colleges of Nursing. (2020). www.aacnnursing.org/Students/Your-Nursing-Career-A-Look-at-the-Facts

Becker, B., Dawson, P., Devine, K., Hannum, C., Hill, S., Leydens, J., Matuskevich, D., Traver, C., & Palmquist, M. (2012). *Case studies*. Writing@CSU. Colorado State University. https://writing.colostate.edu/guides/guide.cfm?guideid=60

Bourdieu, P., & Passeron, J. (1977). Reproduction in education, society, and culture. Sage Publishing.

Bowen, G. A. (2009). Supporting a grounded theory with an audit trail: An illustration. *International Journal of Social Research Methodology*, *12*(4), 305–316. https://doi.org/10.1080/13645570802156196

Cooper, K. M., Ashley, M., & Brownell, S. E. (2017). Using expectancy value theory as a framework to reduce student resistance to active learning: A proof of concept.

Journal of Microbiology and Biology Education, 18(2). https://doi.org/10.1128/jmbe.v18i2.1289

Echeverria-Estrada, C., & Batalova, J. (2019). *Sub-Saharan African immigrants in the United States.* Migration Policy Institute. www.migrationpolicy.org/article/sub-saharan-african-immigrants-united-states-2018

Geertz, C. (1973). *The interpretation of cultures.* Basic Books.

Gist, C. D. (2017). Voices of aspiring teachers of color: Unraveling the double bind in teacher education. *Urban Education, 52*(8), 927–956. https://doi.org/10.1177/0042085915623339

Hannigan, S., Raphael, J., White, P., Bragg, L. A., & Clark, J. C. (2016). Collaborative reflective experience and practice in education explored through self-study and arts-based research. *Creative Approaches to Research, 9*(1), 84–110.

Hauge, K. (2021). Self-study research: Challenges and opportunities in teacher education. In M. J. Hernandez-Serrano (Ed.), *Teacher education in the 21st century: Emerging skills for a changing world* (Chapter 9). Intech Open. www.intechopen.com/chapters/75416

Kim, A. Y., Sinatra, G. M., & Seyranian, V. (2018). Developing a STEM identity among young women: A social identity perspective. *Review of Educational Research, 88*(4), 589–625. www.jstor.org/stable/45277264

Lawrence, S. A., Johnson, T., & Small, C. (2019). Watering our own lawn: Exploring the impact of a collaborative approach to recruiting African Caribbean STEM majors into teaching. *Journal of Negro Education, 88*(3), 391–406. https://doi.org/10.7709/jnegroeducation.88.3.0391

Matsko, K. K., Ronfeldt, M., Nolan, H. G., Klugman, J., Reininger, M., & Brockman, S.L. (2020). Cooperating teacher as model and coach: What leads to student teachers' perceptions of preparedness? *Journal of Teacher Education, 71*(1), 41–62. doi:10.1177/0022487118791992

Mills, A. J., Durepos, G., & Wiebe, E. (2010). Multiple-case designs. In A. J. Mills, G. Durepos, & E. Wiebe (Eds.), *Encyclopedia of case study research.* Sage. https://doi.org/10.4135/9781412957397

Rashid, Y., Rashid, A., Warraich, M. A., Sabir, S. S., & Waseem, A. (2019). Case study method: A step-by-step guide for business researchers. *International Journal of Qualitative Methods, 18*, 1–13. https://doi.org/10.1177/1609406919862424

Rogers-Ard, R., Knaus, C. B., Epstein, K. K., & Mayfield, K. (2012). Racial diversity sounds nice; systems transformation? Not so much: Developing urban teachers of color. Urban Education, 48 (3), 451–479. https://doi.org/10.1177/0042085912454441

Sampras, A. P. (2011). *Self-study teacher research: Improving your practice through collaborative inquiry.* Sage Publishing.

Samuelson, C. C., & Litzler, E. (2016). Community cultural wealth: An assets-based approach to persistence of engineering students of color. *Journal of Engineering Education, 105*(1), 93–117. https://doi.org/10.1002/jee.20110

Stanton-Salazar, R. D. (1997). A social capital framework for understanding the socialization of racial minority children. *Harvard Educational Review, 67*(1), 1–40.

Tight, M. (2017). *Understanding case study research: Small-scale research with meaning.* Sage Publishing.

Wallace, D. L., & Gagen, L. M. (2020). African American males' decisions to teach: Barriers, motivations, and supports necessary for completing a teacher preparation program. *Education and Urban Society*, *52*(3), 415–432. https://doi.org/10.1177/0013124519846294

White, L., & Jarvis, J. (n.d.). *Self-study: A developing research approach for professional learning*. www.herts.ac.uk/link/volume-4,-issue-1/self-study-a-developing-research-approach-for-professional-learning

Zong, J., & Batalova, J. (2019). *Caribbean immigrants in the United States*. Migration Policy Institute. www.migrationpolicy.org/article/caribbean-immigrants-united-states-2017

Appendix A

TIPS teaching interest survey

Survey

Thank you for completing this survey.
The purpose of the survey is to learn more about the career goals of science, technology, engineering, math (STEM) majors and the factors influencing STEM majors' decision when selecting or not selecting a career in Education.
Your feedback will help to inform strategies for recruiting more STEM majors into teaching.

* Required

1. What is your current status? *

   Mark only one oval.

   - ◯ Freshman
   - ◯ Sophomore
   - ◯ Junior
   - ◯ Senior

2. What is your current major? *

   Mark only one oval.

   - ◯ Mathematics
   - ◯ Science / Biology
   - ◯ Other

3. If you have a minor, indicate it here. If you have no minor, skip to the next question.

   ____

   ____

   ____

   ____

   ____

Figure A.1 TIPS Teaching Interest Survey

4. What are your current career goals/ options? *

5. What factors influence your decision to choose or not choose a career in teaching? (Check all that apply)

Check all that apply.

- ☐ Teaching certification exams and requirements
- ☐ (Lack of) Teaching experiences
- ☐ Teachers' Salary
- ☐ Teachers' Schedule
- ☐ Family pressure/ expectations/ demands

6. Please provide more information about the factors you selected above.

7. If we were to become a teacher, what subject(s) would you teach? *

Mark only one oval.

- ◯ Mathematics
- ◯ Science / Biology
- ◯ Other

Figure A.2 TIPS Teaching Interest Survey

8. If you were to become a teacher, what grade level(s) would you want to teach? *

Mark only one oval.

- Elementary (grades 1-6)
- Middle School (grades 6-8)
- High School (grades 9-12)
- Other

9. Have you ever tutored peers or other students in your subject/major? *

Mark only one oval.

- Yes
- No

10. Have you ever attended any teacher education workshops or trainings to learn more about teaching or the profession?

Mark only one oval.

- Yes
- No

11. What topics, information, experiences would be helpful as you consider and explore the optio of becoming a teacher?

Figure A.3 TIPS Teaching Interest Survey

Appendix B

Excerpt from TIPS end-of-year survey

4. What year did you first participate in the TIPS program?:

Mark only one oval.

- 2018
- 2019
- 2020
- Other: ____________

5. Before participating in the TIPS program I planned to pursue:

Mark only one oval.

- Medicine
- Dentistry
- Allied Health
- Pharmacy
- Graduate School-Masters
- Graduate School-Ph.D.
- Nursing
- Psychology
- Social Work
- Unsure
- Other: ____________

Figure B.1 Excerpt from TIPS end-of-year survey

6. Since participating in the TIPS, I have decided not to pursue a healthcare career?

   Mark only one oval.

   ◯ Strongly disagree
   ◯ Disagree
   ◯ Neutral
   ◯ Agree
   ◯ Strongly agree

7. The TIPS program has increased my interest in pursuing a teaching career

   Mark only one oval.

   ◯ Strongly disagree
   ◯ Disagree
   ◯ Neutral
   ◯ Agree
   ◯ Strongly agree

8. I am aware of the challenges students of color face in pursuing STEM disciplines

   Mark only one oval.

   ◯ Strongly disagree
   ◯ Disagree
   ◯ Neutral
   ◯ Agree
   ◯ Strongly agree

Figure B.2 Excerpt from TIPS end-of-year survey

9. I have faced challenges in pursuing a STEM major

   Mark only one oval.

   - Strongly disagree
   - Disagree
   - Neutral
   - Agree
   - Strongly agree

10. The TIPS program has made me more confident in pursuing a teaching career?

    Mark only one oval.

    - Strongly disagree
    - Disagree
    - Neutral
    - Agree
    - Strongly agree

11. My exposure to teaching, professional development and teachers in the classroom has reinforced my decision to pursue a teaching career

    Mark only one oval.

    - Strongly disagree
    - Disagree
    - Neutral
    - Agree
    - Strongly agree

Figure B.3 Excerpt from TIPS end-of-year survey

12. My exposure to the TIPS professional development in summer 2020 caused me to change m mind about pursuing a teaching career

Mark only one oval.

- Strongly disagree
- Disagree
- Neutral
- Agree
- Strongly agree

13. How likely were you to pursue a career in teaching before participating in the TIPS program?

Mark only one oval.

- Strongly disagree
- Disagree
- Neutral
- Agree
- Strongly agree

14. How likely are you to pursue a career in teaching after participating in the TIPS program

Mark only one oval.

1	2	3	4	5
○	○	○	○	○

Figure B.4 Excerpt from TIPS end-of-year survey

15. How satisfied are you with your overall experience in the TIPS program

Mark only one oval.

	1	2	3	4	5	
Very	◯	◯	◯	◯	◯	Very Satisfied

16. My initial career interest in STEM or healthcare was influenced by my family

Mark only one oval.

◯ Strongly disagree
◯ Disagree
◯ Neutral
◯ Agree
◯ Strongly agree

17. My family would respect my pursuit of a teaching career

Mark only one oval.

◯ Strongly disagree
◯ Disagree
◯ Neutral
◯ Agree
◯ Strongly agree

Figure B.5 Excerpt from TIPS end-of-year survey

18. I feel confident about pursuing a teaching career

Mark only one oval.

- Strongly disagree
- Disagree
- Neutral
- Agree
- Strongly agree

19. I am worried that I would disappoint my family or friends by pursuing a teaching career

Mark only one oval.

- Strongly disagree
- Disagree
- Neutral
- Agree
- Strongly agree

20. I am confident that after participating in the TIPS professional development series I understar how to teach STEM content

Mark only one oval.

- Strongly disagree
- Disagree
- Neutral
- Agree
- Strongly agree

Figure B.6 Excerpt from TIPS end-of-year survey

Appendix C

TIPS FUSION professional development honeybee PD (Phase II)

I) Introduction

In the summer of 2020, TIPS and FUSION initiated a very successful collaboration that focused on developing a series of Honeybee-themed Professional Development Workshops. The PD workshops focused on gaining insight from Honeybees in order to enhance teaching and learning. The five Honeybee PD Workshops focused on themes related to (1) Honeybee Flight, (2) Honeybee Stings, (3) Honeybee Honey, (4) Honeybee Communication, and (5) Honeybee Future Prospects and Projects. It has been a wonderful learning journey that continues to expand. We have just touched the tip of the iceberg with our Honeybee Professional Development Series! There is still so much to explore regarding Honeybees and what we can learn from them and with them to enhance and inspire teaching and learning.

The next phase of the TIPS FUSION PD collaboration will focus on empowering teachers to take an active role in designing inquiry-based learning modules and resources that will inspire transformative learning in their classroom. This process will help to enhance their own personal learning and fuel higher levels of inspiration and passion for learning that will flow outward to their students and colleagues.

The approach that is being developed and implemented for Honeybee PD is to co-develop a comprehensive system of analogies, coaching, mentoring, assessment and evaluation, and entrepreneurial resources. These resources will be instrumental in creating multidimensional PD that focuses on teamwork and building fruitful learning pathways that connect teachers' and students' areas of passion with essential lessons for life.

The teamwork aspect of Honeybee PD will be integrated into a larger framework called Teaching Excellence by Analogy and Mentoring (TEAM). Bringing more inspiration and passion into teaching and learning will be facilitated by expanding the field and reach of PD through creating a comprehensive PD platform for the design, implementation, and assessment/evaluation of PD called Personal and Professional Passion Development by Dynamic Design and Psychological Discovery (PD). The overall vision for TEAM, PD focuses on developing a learning experience where teachers and students co-design an intellectually stimulating environment full of passion, excitement, and learning "buzz." Teachers and student scholars are empowered to be confident and believe in themselves and their developing abilities. They ask wise and insightful questions and explore and grow while collecting resources to bring back and share. Through teaching and learning collectively as a team, a much more valuable and powerful learning resource will be produced to support the team as they learn together.

II) Goals and objectives

The overall questions and objectives for this project are listed below.

1. How can we most effectively learn from and model key aspects of Honeybee life, physiology, and psychology to transform teaching and learning in our schools? We will expand upon our initial lessons through guided activities and learning opportunities that will challenge and help our TIPS Teacher Cohort develop skills, tools, and resources that will propel our students and teachers to fly with purpose and impact as they explore the STEM/STREAM Universe and learn at full capacity. We will draw upon and build further inspiration based on Honeybee lessons related to flight (Flight Conduction), sting (Knowledge Transduction), honey (Honey Production), wax (Wax Construction), and communication (Dance Fluxion).
2. What insights can we gain from Honeybees to learn how to make education equivalent of honey? Honeybees use abundant teamwork to gather a low-value raw resource (nectar) and process it to create a higher-value energy-dense product.
3. How do we best equip TIPS scholars with the inspiration and skills to shift from being PD consumers to being PD creators and entrepreneurs? How can we work together to produce a marketable product that contributes to the health of the hive by providing rich resources to facilitate learning opportunities?

There are a number of fundamental ideas that will drive the development of our Honeybee PD Phase II Project. Building on the themes of teamwork, passion, and TEAM and PD, we will work together like Honeybees to collect PD resources, bring them back to our learning hive, and process them into rich value-added resources that we will use to nurture our young Honeybee scholars in the classroom. We also look forward to helping our teachers' progress through the stages of PD interactions. Teachers begin their learning journey as PD consumers, and one of our program goals is to design learning pathways to help teachers progress to become PD Contributors, PD Curators, PD Designers (Artisan/Artists), and PD Creators.

There are a number of examples that highlight the benefits that can be derived from our Phase II efforts. While Honeybees provide the central theme for our engagement, the ideas behind TEAM PD allow flexibility for exploration in other areas that may or may not be related to Honeybees. For example, one way of framing TEAM involves developing visual analogies that encompass various aspects of what teachers can do and become in exploring unique learning pathways. Some examples related to Honeybees and other topics include:

1) Teacher as Honeybee—While this is highlighted in our current program, there are many more resources that we can develop based on our Honeybee theme.
2) Teacher as Fisher—What are the best inspired ideas to attract students to engage in learning?
3) Teacher as Miner/Prospector—How do we help students develop the skills to search for learning patiently and diligently to find gold nuggets or that gem of great worth?
4) Teacher as Materials Scientist—How can teachers and students be exposed to the best ideas and cutting-edge developments that are being uncovered? A few weeks ago, a long-sought-after material was discovered in the form of superconducting materials that function at room temperature. There are a lot of materials science and biomaterials science analogies that can be developed to enrich STREAM learning. One in particular that we can explore involves the concept of functionally gradient materials. This topic was introduced in our initial exploration of the Honeybee sting. The stinger is softer at the tip and gets harder along its length. This structural adaptation promotes more efficient penetration of the skin for the delivery of the bee venom. We can explore functionally graded learning as a PD resource concept where we model and practice functionally graded transition to help scaffold learning in the best manner for optimized learning for teachers and students.

III) Program structure

Each month, a theme will serve as the focal point. We will have one group team activity monthly and one self-directed PD project activity that can be built over multiple months. The self-directed PD project will be similar to the Library of Congress TPS project write-up that will be constructed around a deep key question. The sample questions that I often ask include: "How does a bee learn how to fly/build the hive/dance and communicate information? How does the Honeybee brain work? Why do they do what they do? Can we put a camera on Honeybees to see inside the hive?" Our teachers will select one PD topic for designing a series of integrated Learning Journey Pathway PD resources to help teachers take an inspired journey in transitioning from beginner to master teacher/learning coach. We will adapt the Functionally Graded Transition (FGT) Model for PD Design and Creativity for use with our TIPS scholars. We will walk through this PD resource design process in detail, including modeling the process of functional grading and staged transitions in learning and scaffolding to construct the best learning experience possible.

1) Honeybee flight (lift and conduction)

The dream to achieve flight has been in people's hearts since we observed Honeybees and other insects and birds that have the ability to fly. While we do not possess the innate resources for unassisted flight, that did not prevent countless explorers, innovators, and entrepreneurs who studied, asked the right key questions, and eventually succeeded in achieving flight and human-powered flight. Flight and the dream of flying provide a number of wonderful analogies for scaffolding rich learning, and this will serve as one of the learning threads that we will explore. One activity that we can build involves using the foldscope to observe bee wing structure.

2) Honeybee sting (knowledge transfer and transduction)

In Phase I of our studies, we learned about honeybee stings. While the experience was not pleasant, we learned how it is designed to deliver maximum pain and irritation. On the reverse side, the sting has served as a source of inspiration in terms of Mikala and her entrepreneurial enterprise and empire related to honey-sweetened lemonade. She started this as a child, and with the right level of encouragement and support, she has transformed it into a multimillion-dollar outreach that focuses on honeybee education. This is quite a goal to achieve by the age of 15, in addition to being a published author. How can we work with our teachers and students to "sting" them with inspiration and put even bigger dreams in their minds and hearts during Phase II? We will work together to develop more powerful learning analogies and resources that result in more powerful knowledge transfer and transduction to build our learning hive in the classroom.

3) Honey (energy production)

Each honeybee plays a key role in making honey for the hive. This team effort has serious implications in terms of the need for everyone to make a contribution to achieve the best outcome for all. The amount of nectar contributed by each honeybee on an individual basis may seem miniscule, but each honeybee contributes what they can collect and gather, and the end result from the community turns into great stores of honey that provide precious nutrition for the long winter. Honeybees also make food products such as BeeBread, propolis, and Royal Jelly.

Another fascinating aspect of honey involves the mechanism by which it is transformed into wax for building the hive. How does this happen? What are the energetics of this conversion process, and what can we learn from this?

4) Wax (resource construction)

Wax is a wonderful, versatile building material that is created by honeybees via the chemical conversion and biological transformation of honey. Wax is literally exuded as thin plates from spaces on the honeybees' abdomen. Honeybees work together to collect the wax, chew it, and use this compound as the structural foundation for the hive. At a fundamental level, people are builders, creators, dreamers, inspirers. We do our best work when we are inspired and work together to inspire others.

We will focus on designing simple yet elegant learning building blocks from which we can build PD resources for teachers and their classroom learning hive. We will learn from traditional building resources (modeling clay, construction toys (Legos, Kinex) and newer resource kits (STEM/STREAM robotics, magnetic construction)) to create additional learning resources. A few examples include (1) origami and modular origami using post-it notes, (2) Hexastick models (pencils, wooden rods, toothpicks), and (3) Spragnets. The most compelling aspect of this portion of Honeybee PD involves the lessons that we can learn from our Honeybee partners in terms of our ability to take very simple building blocks (paper, pencil, springs, magnets) and transform them into inspiring structures that stimulate neural connections and deeper thoughts and ideas based on enhanced visualization and perception. The construction model kits will be designed to be age-appropriate and bring joy to the classroom via connecting hands-on learning with fascinating concepts. Bendaroos and wikisticks are trademarks for wax-covered strings that can be used to build impressive structures that are safe for children as young as pre-K. Pipe cleaners can be used at the elementary school level to add additional variety to the other geometric model building resources.

Higher-level model building and construction resources can be developed for middle and high school scholars as we work to equip teachers with the resources and content knowledge to help their students stretch to achieve higher levels of inspired learning in the classroom and beyond. Examples include individual and team-based models and projects that can develop tools to remotely observe the secret life of honeybees and citizen science projects that explore colony hive health to address causes for colony collapse disorder (CCD). In 2019, an undergraduate Student Design Team at Catholic University designed a system of sensors to collect continuous temperature measurements in real time to monitor hive health. This is an example of a project that could easily be adjusted and scaled to spark inspiration and innovation in young scholars at the elementary, middle, and high school levels.

5) Communication (dance fluxion)

Fluxion is a word that encompasses the action of flowing and changing. It has a surprising etymology in that it was coined by Sir Isaac Newton as he was studying rates of change for studying stars and planets while developing the foundations of the calculus. There are a plethora of additional connections between honeybees and various content areas in the STREAM Universe, including mathematics. There are some fascinating honeybee facts related to why the hexagonal shape is favored in hive construction, with architectural, physics, chemistry, geometry (tiling), and calculus insights involved.

Appendix C: Examples of models shared during PD workshops

Figure C.1 Examples of models shared during PD workshops

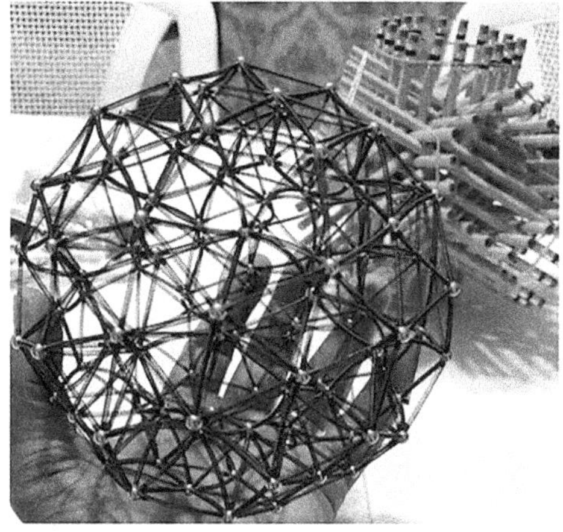

Figure C.2 Examples of models shared during PD workshops

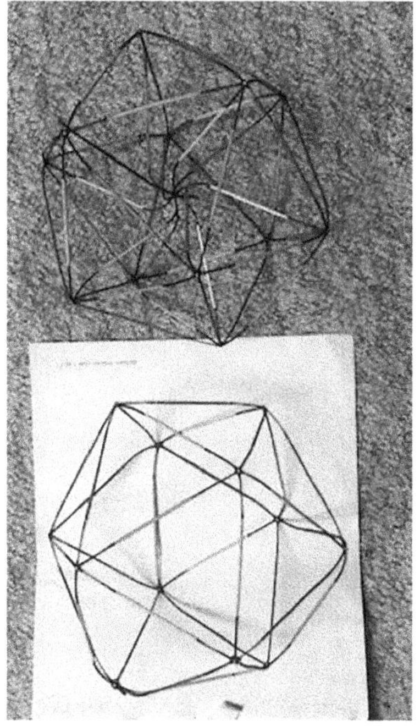

Figure C.3 Examples of models shared during PD workshops

Index

Note: Page numbers in *italics* indicate a figure and page numbers in **bold** indicate a table on the corresponding page.

For Product Safety Concerns and Information please contact our EU representative GPSR@taylorandfrancis.com
Taylor & Francis Verlag GmbH, Kaufingerstraße 24, 80331 München, Germany

www.ingramcontent.com/pod-product-compliance
Lightning Source LLC
Chambersburg PA
CBHW061828220326
41599CB00027B/5224

* 9 7 8 1 0 3 2 5 3 9 5 1 5 *